INFLUENCE OF MUSIC

ON

HEALTH AND LIFE.

PART I.

CHAPTER I.

General reflections on the melodic and harmonic character of music among different nations.—Influence of the tone of the voice.

EVERY nation has a music and a song of its own. All, without exception, according to their taste, their talent, and their customs, have made use of the organs with which they have been endowed by nature, and of the instruments invented by their own ingenuity.

Among primitive nations, music, having neither rules nor laws, consisted only of melody, which is a succession of sounds gratifying to the ear; it is a pleasure of emotion and of sentiment, addressed to the heart.

Harmony is a music of science and combination. It is the result of several sounds heard at once, which produce what we call *accord.* It affords a pleasurable sensation which is addressed to the understanding.

As to rhythm, one of the three elements which are found in nature, it pervades all melody and imparts character to it. Martianus Capella, a compiler of the fifth century, wrote at that time: "Melody is the female, rhythm the male." For us, melody is the sketch, harmony the coloring, and rhythm the character."

Music is not the same among all nations. Climate, customs, habits, a certain degree of refinement or of civilization, all impress the tones of melody with various characters.

Northern nations express themselves very differently from those of the south, while eastern nations have quite another music from those of the west.

The exposition of 1867, that immense caravansary, where all nations elbowed each other, that vast *capharnöum* where the most diverse objects were crowded together pell-mell, has afforded us samples of the music of many nations. To our ears the Chinese music, although executed by Parisian artists, seemed nothing but barbarous discord.

The Arabian music, harsh and nasal, made our very flesh creep.

The character of the music of any nation, its manner of producing tones, of sustaining, of seizing, and of dropping them, as well as the intonations and inflections of the voice, can, up to a certain point, give an idea of its intelligence, manners, and customs. They are the thermometer of its physical sensitiveness and moral sentiments.

The degree of civilization in a nation, the various epochs of its growth, the kind of government which it has developed, or which has been imposed upon it, the manner in which it is conducted, and its social conditions, are so many causes for varieties of melodic character.

Savage nations have but one kind of music, that of nature. Civilized nations have two: the popular or national, which always preserves (in spite of all the changes wrought upon it by science) a certain flavor of the national *soil* (if I may be allowed this expression), and the music of the so-called cultivated circles. One is simple or lively, gay, melancholy, or dreamy, just in accordance with the character of the nation: the other on the contrary is studied and formal, soft and tender, or stiff and harsh, according to the ways and habits of those among whom it is cultivated. Compare the popular songs of Brittany, of the Vosges, of the Pyrenees, of Auvergne, with those of Paris, and you will be struck by the difference of the melodies. The inhabitants of mountainous districts have a different accent from those living in level countries, and the Danes, Swedes, and Russians do not sing like the Italians, Spaniards, or Portuguese. Education, no matter how powerfully it may affect the melody of a people, can never quite destroy its national character. In Italy, for instance, in that classical land of musical science, where harmonic combinations have been studied and used, with infinite skill, where all the inhabitants, of every class and of every degree of cultivation, sing the *Arias* of Rossini, Bellini, Donizetti, and Verdi, the popular songs, even in the various provinces, are all marked by peculiar characteristics. The popular song of Rome differs as much from that of Naples, as the Canzonetta of Venice differs from that of Florence; but the *saltarellos*, *tarantelles*, *barcarolles*, songs which are essentially *national*, always preserve a certain wild and fascinating originality.

The influence of climate upon the tones of melody is peculiar, imparting to them a certain individual character, which never varies. Music expressive of languor and of love, of sweet and tender melancholy, belongs exclusively to warm climates; cold climates inspire music expressive of martial ardor, and of conquest, while lively and graceful music is heard mostly in temperate climates.

The Frenchman, gay, witty, and capricious, when he can abandon himself to the emotions of his happy nature; when not wholly absorbed by the preoccupations of his material and social surroundings; when not crushed by the hand of tyranny or despotism; when, in short, he finds himself in the full enjoyment of his rights and liberties—the Frenchman loves his merry song. His melodies are short, airy, and gay. The German, on the contrary, calm, phlegmatic, speculative, gives to his phrases more vagueness, more poesy, more of the ideal. His *Lieder* are stamped with a shade of reverential melancholy, and symphonies are works for which he has a special predilection. "The Italians," said Grétry, "breathe forth their melodies with a sigh, because they feel too keenly: the Germans sigh for the voluptuous sensations, for which they long."

The Englishman sings as he lives; he is positive, stubborn, and formal. His melodies are as poetical as a coin of a hundred sous, passionate and exciting as his steam engines, cheerful as his own dark leaden sky, clear and flowing as the waters of the Thames. When he sings, not a smile ever graces his lips, not the slightest contraction disturbs the fixedness of his countenance, not even a spark of enthusiasm accelerates the beating of his heart.

If, then, it were only granted us to know the character of the music, above all, of the popular or national songs of the races who have preceded us, it would be easy for us to reproduce their history, to discover their habits and customs, to apprehend their character, and to form an estimate of their intelligence, their dispositions, their natural talent for fine arts and the sciences; in short, to appreciate the general bent of their minds.

If we pass from these general reflections to particular ones, we shall remark that every living species is gifted with a voice and song peculiar to itself. The voice, the direct means of disclosing one's inmost thoughts, deepest sentiments, and warmest affections, is capable of the greatest variety of tone among the human species. Nevertheless, no matter how great the variations may be, the tone of the voice always gives expression to the sensations, emotions, and passions.

As all the world knows, voices are sharp, deep, true, false, clear or muffled; but spite of these varieties, the human voice has its own peculiar, individual character. Man can indeed, up to a certain point, alter his voice or imitate that of a majority of animals; but, however much he may study, however great may be his perseverance and his power of imitation, he will never succeed in completely changing it. It is the same with animals. The parrot, the magpie, and a few other birds, can indeed, after patient teaching, be made to imitate speech and the human voice, but never so perfectly as to deceive a practised ear. In the human species, the imitation of the tone and inflection of the voice of one individual by another, is often very remarkable and quite unstudied. Two persons, a man and woman for instance, living constantly together, can adopt each other's intonations, but these same intonations, creating such surprise in ordinary conversation, will disappear very suddenly, if expression is to be given to the passions or any deep emotions; at such times, every one resumes the tone of voice which belongs to him, individually. Nature preserves its originality, at all times, under all circumstances. The seal which she has imprinted upon every being, is never so thoroughly effaced that the individuality of any one can be permanently lost.

Among animals, the tone and variety of voice are manifested in a manner still more marked than is the case with the human species. With them the tone and diversity of voice are so extraordinary, so marvellous even, that the sensations and emotions with which they inspire us are unspeakable and indescribable. Let us picture to ourselves, for a moment, a man whose powers are all fully developed, except that he has never heard or conceived of any sound, and fancy him placed all at once in the midst of numerous animals of different species, all of whom are made to sing or scream, and then that he is suddenly gifted with the sense of hearing; what sensations would he not experience in listening for the first time to the confused noise which now suddenly strikes his ear? With what terror would he not be inspired at the bellowing of the bull, the neighing of the horse, and even the braying of the innocent donkey? With what sweet emotion would he not be filled at the melodious song of the linnet or the nightingale? How could he ever endure, in short, the sound of all the voices to which we have become gradually accustomed, either by habit or education. It would require much time, as well as close application, before he could even distinguish the voice of each species. The pig grunts, the lion roars, the dove coos, the frog croaks, the dog barks, the cat mews, the raven caws, the sheep bleats; birds do not hiss like serpents, the canary and the goldfinch do not sing like the nightingale.

The variety and diversity of voices has been a subject of special interest to a large number of students. By the voice, the song of animals, the various species can be recognized and distinguished; according to the tone of the human voice it has been deemed possible even to determine the moral character of an individual, his tastes, instincts, and passions. Could this study be based on fixed principles, on irrefutable observation, it might have the happiest influence on our action in life, and our social relations, but unfortunately, up to the present time it remains a mere theory, simply a subject of vague conjecture, about which *savants*, even those of the earliest ages, have always maintained fruitless discussion.

Plato, who passed his whole life in study, investigation, and teaching, had the habit, we are told, of making any man whom it was his interest to know well, read or talk with him in a loud voice. The quality of the voice, the intonations, the inflections of the speaker or reader, were to the philosopher so many indications, if not certain, at least probable ones, of his moral character. And, in our own time, Grétry, in his "Essays on Music," asserts that he has never been mistaken in the opinion he has formed of individuals who had said nothing more to him than "Good-day, sir," or "Good-morning, my friend." According to the intonation with which these words were spoken, the great musician assumed to guess with whom he had to deal, and regulated his conduct in accordance with the impression he had thereby received.

"A 'Good-morning' is almost always sufficient to enable me," said he, "to appreciate in general the pretension or the simplicity of a man.

In conversation a man often hides his real character from us, either through politeness or duplicity, but he has not quite learned how to disguise the intonations of his voice. This 'Good-day, sir,' and 'Good-morning, my friend,' put to music with their exact intonations, would show what a power vanity is, and how quickly the key changes when its influence ceases to be the ruling one."

One need not be Plato or Grétry, indeed, to form some estimate of an individual by the intonations of his voice, and it is above all, in the present age of versatility and prejudice, of humility and arrogance, of ambition and servility, of self depreciation and vanity, of impudence and modesty, of timidity and boldness, that a "good-day, sir," or "Good-morning, my friend," may enable us to fathom the very soul of him who utters it. It is the quality of tone that makes the music, according to an old saying, and "sir," "my friend," "yes," "no," spoken in a certain manner, often express more to us than a hundred words.

It is not only of the intentions of people that judgment has been formed from merely listening to the tones of the voice, but it is even assumed that their tastes, their instincts, and inclinations may be thus discovered. In order to attain this result, rules have been given which are said to be invariable.

A clergyman by the name of Kircher has written a folio in Latin on this subject, and the book is as confusing to read as it is heavy to carry. Its title is "Musurgia universalis." In it he says (remember that I neither claim to agree with, nor be responsible for what I am about to repeat, for reverend gentlemen of ancient times, not unlike those of our own, often gave utterance to words which cannot be accepted as *gospel-truth*): "Those who have a deep sonorous voice, like that of a donkey (excuse the comparison) are indiscreet, and quarrelsome." According to the original. "*Asinus admodum magnam vocem habet et gravem, ut asinus indiscretus, ut petulans et contumeliosus; ergo, quorum magna et gravis vox est, illi sunt petulantes, indiscreti, contumeliosi.*" "Those who have a sharp, thin, and husky voice, are weak, effeminate, and easily yield to low passions."

"A full, abrupt voice denotes a strong, impulsive, bold, enterprising man.

"The voice whose tone is weak, beside being sharp and drawling, gives evidence of a character lacking in energy and firmness; so, persons gifted with such a voice are timid, patient, having no will of their own, no marked individuality. Not sensitive to injuries, they preserve no recollection of them, and never seek revenge.

"Men with a powerful, deep voice are generally cowardly. Haughty with their inferiors, they are insolent in prosperity, but faint-hearted in adversity. So it was with Caligula, according to Tacitus.

"Those individuals whose voice, deep at first, is raised to a high key as they finish speaking, are noisy, irritable, and of unhappy dispositions.

"Lastly. Those whose voice is thin and shrill, are peevish, ill-tempered, passionate, possessing characteristics closely resembling those of a goat." Have these remarks of Kircher all the worth with which their author attempts to endow them? Could the surroundings in which he lived have furnished enough examples to enable him to establish his theory; could the great and small passions which hold their sway in the cloister and in the religious world, or the jealous rivalries which are kindled by the ardor of faith, or the hatred often hidden by humility of demeanor, of looks, or of words, have given fit opportunity to the reverend father for the remarks and principles that he has here put forth? This I am not prepared to say. As to ourselves, since our attention was first fixed on this point, it has not been possible for us to form an opinion of any value, in the matter. Let us rather allow each one the privilege of verifying for himself the exactness of the facts which have just been related, and for the truth of which we are not prepared to vouch. If now we are to study the ton of the voice among individuals, for the sake of drawing our deductions and anticipating consequences, we must consider man when in a serene state of body and mind; otherwise we should risk making serious mistakes, and committing grave errors. We know well all the changes and modifications of which the voice is capable, when the passions are roused, or the soul is deeply moved.

Fear and weakness cause it to lower, astonishment makes it abrupt, admiration prolongs it, anger is expressed in hoarse tones, joy in full, rich ones, hope speaks in a firm, even voice, desire, in hurried events, and long exclamations mark the beginning of conversation.

In a time happily long before our own, in an age which to the honor of all humanity will never return, the voice of those wretches destined to become singers of the Sistine Chapel was changed by mutilation of the human body. The Popes, representatives of a good and merciful God, did not shudder at a crime so odious that it might be called treason to all that the word *humanity* suggests, and if they did not encourage, at least countenanced human sacrifices which might have even roused the jealousy of Pagan gods. The unhappy victims did not all die; but mutilated, rendered effeminate, condemned henceforth to know no joy, no pleasure, they were destined to lead a sad and desolate life, to which death would have been preferable. Time and civilization have happily visited justice on such crimes, and if any such mutilated beings are still to be found in the papal choir, it is only through fraud, or owing to the ambition of unnatural parents.

The voice, the bond of union in the social and political world, and the direct means of expressing thought, has an important influence on all the relations of life, and is all-powerful in moving large assemblies. Eloquence, according to many authors, whose opinion I share unreservedly, is based on the quality of this vocal organ. The most eloquent orator is, then, supposing this idea to be the true one, he who speaks in the most fitting tones, musically considered.

J. J. Rousseau has recently lent the support of his opinion to the same view. He says, "The diverse sensations which we experience from music and eloquence, which is only *spoken* music, or recitative (if I may be allowed this expression), depend upon a multitude of causes; their source may be traced to an infinite variety of reasons; one of them, the most powerful, and the most irrefutable, is, without doubt, the agreeable concord of the harmony. Do we not all recognize what an exquisite pleasure the ear receives from a consonance, while dissonances seem to grate, as it were, on the tympanum, and affect it painfully. Now, if agreeable sounds, followed by a series of other agreeable sounds, have the power of pleasing any sense—that of hearing, for instance—the pleasure results from this succession of concordant sounds. Why is it thus? It is as difficult to tell as it would be to say why rose color is more pleasing than gray, and why the odor of vanilla is more agreeable than that of the marigold."

"If, then, an orator addresses his audience in discordant tones, far from producing gratification, from pleasing or fascinating them, and rousing them to enthusiasm, he will only inspire them with uncomfortable, even painful sensations; he will repel rather than attract them."

But a *fine* voice is not the only requisite for an orator. According to some authors the voice need never be raised more than a fifth in order to produce excitement; its compass would be limited, then to the distance, for instance, from C to G. Each one of these notes would have special fitness for language or conversation.

The C would be appropriate for explanations.

The D for the clear utterance of vowel sounds.

The E for the tender passions.

The F for arousing to action.

The G for expressing all that pertains to the pathetic.

Had these observations been known and applied in ancient Greece, among the flute-players whose duty it was to accompany or support the orators, would they have served to define to them always the most fitting tone for their harangues? Numerous facts lead me to believe so. In Italy—that classical land of improvisation, of poesy and music—I have often heard *improvisatore*, not such as speak in the open air and in public places, but those really worthy of the name, accompanied on an instrument, played either by themselves or others, the modulations being all within the interval of a fifth; and in this connection I recall an incident which ought to be mentioned here: At a meeting of the Tiberian Academy in Rome, a young woman of equal talent and modesty, Rosa Taddei, improvised for a whole evening on various subjects furnished from among her numerous listeners. This young woman directed the artist who was to accompany her on the piano, as to what kind of modulation, and in what rhythm to play, in order to render music that would be most fitting to the style and character of the subject. The modulations never went above the fifth; the accompaniment was as appropriate to the subject and rhythm of the verses, as if it originated from the same source. Sometimes tender and melancholy, sometimes angry and impassioned, the young *poetessa* fascinated her audience, and received applause justly due to her wonderful talent. The influence of accompaniments on inspiration, and on the effects which can be produced by music, is of immense importance. Therefore all true masters of art spare no effort in the treatment of this powerful auxiliary.

Grétry, whose competency to judge on the subject is undeniable, gives, in his "Essays on Music," some advice which is not only very useful to professional musicians, but instructive and full of interest to amateurs and all appreciative lovers of music.

"The composer," he says, "will produce his work with such shading as may be fitting to his auditorium, but his heart will always prompt the most natural melodies. Without paying much attention to details he will build the structure of his composition on a broad foundation; he will base his melody and harmony on the groundwork of human emotions; by a variety of shading, he will understand how to show the difference between gratified and expectant, or jealous and trustful love. The love of money will not be expressed by the same sounds as the love of dissipation, and the love for the creature will be breathed forth in other accents from those used to express love of the Creator."

"These various shades of the same passion form what might be called the details, and the more closely the composer is able to approach truth in the expression of the various sentiments, so much the nearer will he be to perfection.

"The success of a musical composition depends upon this truthfulness of touch, if I may be allowed the expression. Upon this truthfulness, too, may be said to depend all the sentiments and emotions which are experienced by the audience. The nearer his composition approaches nature, the more his audience will be excited and fascinated.

"A young girl," continues Grétry, "may assure her mother that she has never yet felt love; but while she affects indifference through the means of a simple, monotonous song, the orchestra may express the distress of her already enamored heart. A jealous lover makes long pauses between his strains; the more he fears to accuse his beloved wrongfully, the more emphatically he utters his proud and haughty tones."

"He often uses the chromatic scale, which is sad and at the same time sinister: sad for the jealous man who seeks to arouse our interest, and suggestive of misfortune for those who listen to him, and above all for her who is the object of his jealous suspicion. In general, the sentiment should be made known only through the singing; the action and general character should be expressed by the accompaniments" (Grétry, Essais sur la Musique, tome ii., p. 14).

This advice, essentially practical, has been and is still often followed by many composers; in the overture to "Francs-Juges" by Berlioz, this energetic and fanciful musician, too little appreciated, and far too much criticised, has the violins go through a series of chromatic scales, ascending and descending with startling effect; it is almost like hearing the very sobs and groans of the victims. The shading in accompaniments is not wholly due to inspiration, it is the result of science; therefore the ancients, who found the art of music an unbroken field, marched with hesitating steps.

Uncertain, and groping, as it were, they increased and diminished, hurried or slackened, the sounds which they produced by means of their voice or their instruments, just according to feeling. Therefore their accompaniments are marked by extreme simplicity; they are often lacking in power and force. Our modern composers, on the contrary, keeping pace with the science as it is day by day brought nearer perfection, obtain from the treatises on harmony and orchestration, rules and precepts which they enlarge and enrich by the power of their own genius. The effects which are thus produced, often astonish and sometimes excite us. Of what effort has the human mind not been capable? How many attempts and failures has it not passed through before it succeeded in creating these scientific, though not always agreeable combinations? It is by investigating from the earliest beginnings, that we shall recognize all the phases, all the transformations, all the metamorphoses of this sublime and admirable art of combining sounds. Let us then cast a rapid glance at the origin and history of music.

CHAPTER II.

History of music.—Its character as depending on the physical and moral condition of the races.

IT seems to me impossible to assign a positive and settled origin to music. This art, like all other arts in which the ideal plays the principal rôle, can have no fixed beginning.

As soon as man perceived that he could utter sounds, and impart different intonations and shadings to his voice, he might be expected, according to the circumstances under which he was placed, and the passions with which he was moved, to create a music. It was, without doubt, irregular, without form, not governed by any laws, more or less harmonious, and having melody, fanciful or wild—but still, it existed.

Full of emotion rather than thought, and in this respect differing from the other arts, music springs not only from the deep and inmost sentiments that touch our soul, but from our love for imitating nature.

When two men were under happy influences, when a man and woman drawn together by a natural love of pleasure, could express by the tones of the voice or by language the feelings of their soul, they certainly gave utterance to a series of sounds, which, though the individuals were unconscious of it, formed a melody, that is a song. These sounds, and this language, were more or less harmonious, more or less rhythmical.

Attentive to all sounds, observers listened to the harmonies of nature, and attempted to imitate them. The rustling of the wind in the woods, the murmuring of the brook, the roaring of the tempest, exhaustless sources of sentiments and emotions, furnished endless material. The excitement of feeling, the exaltation of thought, the rapid movement of instinct and passion, all gave evidence of the effects produced by harmony, or by numerous and varied tones. Constantly analyzing with more care his feelings and emotions, man sought their cause, and being naturally given to imitation, he attempted to utter by means of his voice or instruments, though very imperfectly, no doubt, the sounds and noises that had reached his ear.

In his analysis he brought reflection and memory to his aid, and recalling his inmost and sweetest emotions, endeavored to reproduce the sounds that had brought them about; therefore, I have no doubt but that his first songs were those of pleasure, sorrow, happiness, and love. Until then he had been answerable only to his own conscience, so to speak, but soon becoming less absorbed in himself, and growing forgetful of his individuality, he became aware of all his surroundings. He raised his eyes to Heaven; the sun, the moon, the stars, the glorious firmament, the harmony of nature, and its countless sounds, filled him with reflections. Plunged into a delicious melancholy, blessed with a calm spirit, leading a tranquil and quiet life, he asked himself if all that he saw had been created especially for him, and what powerful mind had ordained this gorgeous sight which he now witnessed. He worshipped what he could not understand; he bowed down before the star that he saw appear and disappear on the horizon at fixed times ; the idea, then, of a power greater than his own first crept into his soul, and inspired him with songs of wonder and admiration, of gratitude and veneration. He sang religious airs before even having a religion.

This first period of calmness and repose could not last long; the trials and struggles of life soon took the place of this childlike tranquillity which amounted almost to indifference. The passions were developed, and at the same time rivalry, jealousy and hatred.

People were formed into tribes or clans, and all the passions sprang up together, simultaneously. Then music, its nature closely allied to that of human emotions, speaking directly to the sentiments and thoughts, touching and exciting the heart, animating the spirit—the reflection of all that we understand by *human nature*—assumed new forms.

Certainly, if it were possible to restore the music of races as it was at the time of their earliest beginnings, it would be easy for us to write their history, to ascertain what were their moral faculties, the tendencies of their mind, their character, their manners and customs. It would be easier for us to form this estimate, than if we had to treat of free and independent nations. In fact, the greater the liberty enjoyed by a people, the more character does their song have. Music, at all times and in all countries, bears the seal of independence, servitude, misery, or happiness, according to the condition of the nation which cultivates it. Among people pressed by want, exposed to the horrors of famine or destitution, or surrounded by enemies, songs of war, vengeance, and extermination, escape from throbbing breasts and hearts embittered by misfortune; then the tones are wild, savage, harsh and fierce. They are the fearful expression of passions excited by distress or misery.

The tones of sorrow, grief and remorse, are heard later; for sadness and remorse presuppose the development of moral faculties, of affectional emotions, of friendship, fidelity and devotion; in short, a relative civilization. Can nations, whose very existence is precarious, who are daily forced to put forth every exertion to ward off ever-threatening destruction, experience gentle and tender emotions? I think not.

We should bear in mind that what has just been said of the growth of primitive music is based on no authentic fact, document, or tradition. We must proceed by inference and reflection, for no spark has shed the smallest light on the obscurity and darkness of the artistic life of the early nations, that is, of nations devoid of civilization.

Little by little we are approaching the time when we shall be able to restore some traditions, some fragments, almost obscured it is true, but still easily enough recognized, to enable us to advance with more certainty in our investigations and to discover some vestiges of the art which forms the subject of our work.

If we wish to study the origin of the arts and sciences, we must look toward the East. There, palpable traces of some kind may be found, pertaining to an art which cannot have had any special inventor, since it is, as I have said, the result of the imitation and observation of physical phenomena, which every one can understand.

No one doubts that the people of India, Persia, and China were the first to establish rules for the language of music, and to determine the series of sounds which constitute the scale. A Chinese legend, charmingly related by a critic, Scudo, whose works are full of learning, and give evidence of keen observation, will show how this musical scale was formed.

"Under the reign of I know not what emperor, who lived B. C. 2600," says this amiable critic. "the prime-minister was ordered to put an end to the confusion existing in the musical scale. Obedient to his master, the minister went upon a high mountain which was covered with a bamboo forest. He took one of the canes, cut it between two knots, took out the marrow that filled it, and blowing in the reed thus emptied, a sound came forth which was neither higher nor lower than the tone which was natural to him *when he spoke without being agitated by any passion.*"

"Thus the generating sound of the series was determined. While the minister pursued other experiments necessary to attain the end he had in view, a couple of birds, male and female, came and perched on a neighboring tree. The male began to sing, and uttered six sounds; the female, when replying to him, was heard to utter six others, and it happened that the twelve sounds formed together the twelve degrees of the chromatic scale. The minister, profiting by the lesson he had just received, cut twelve canes of a length necessary to produce the twelve semitones, or chromatic degrees which are contained in an octave."

This fable, a charming fiction, which concerns the theory of music as well as the physical construction of the scale of sounds, contains fundamental truth, confirmed later by searching experiments.

Pythagoras, in ancient times, pursuing, with his well-known zeal and persistence, the idea with which he was entirely absorbed, that of reducing all human knowledge to a mathematical basis, subjected the phenomena of sonorous bodies to a strict calculation. He calculated the number and relative value of the vibrations, and established the absolute correctness of the intervals contained within the limits of an octave, which led Leibnitz to say, later, "Music is a calculation which the soul makes unconsciously, in secret."

The Egyptians, Hebrews and Jews have left us positive and unmistakable proofs of the existence of a music governed by rules and subject to fixed laws and principles.

The melodies of these nations have not come down to us, but the sculptures on their monuments, the representation of some of their instruments cut on their propylæa and obelisks are, no doubt, physical and palpable proofs of the existence of a music. This music, in our opinion, must have been well regulated, because the Egyptians, who were keen observers, eminent as calculators, skilled in astronomy and mathematics, could not have helped giving a rhythm and measure to the art, making laws, establishing fixed and definite principles, or even writing treatises on harmony. Unfortunately we possess no written record, but here conjecture amounts almost to certainty.

The Greeks, disciples and imitators of the Egyptians, followed for a very long time the traditions handed down by their predecessors. Their music, simple melody, presents nothing new, startling, or original. Considered as a mere embellishment of poetry, as a help to language, as a guide to orators, rather than as an absolute and independent science, it remained stationary. But very soon the luxury of the people, their tendency to the production of literary work, their brilliant imagination always ready to imprint a poetic tinge on all the subjects which they considered or discussed, made them shake off, in their efforts, the yoke of tradition and routine. Starting from this grand and noble principle, that all the arts tend to soften and perfect the manners of a people, the Greeks, sensitive to the harmony of language, and to the rhythm of speech, turned their attention to poetry and eloquence. Measure and accent in poetry, like rhythm in eloquence, which can be accelerated or slackened at the will of the poet, and according to the expression of the idea and the character of the passions, were regulated by the order of the syllables, and admirably aided a language, the euphony of which was its greatest charm. Poetry, which is already harmony, was to receive a new impetus from the powerful auxiliary, music.

The philosophers, and they were numerous, were the first to profit by the resources at hand, in melody and rhythm, in order to set forth their ideas in a more imposing and seductive manner. Wishing to harmonize the intellect and faculties of the soul, by fascinating the senses, they made use of the means above-mentioned for inciting the heart to laudable action, and for implanting there germs of morality.

Noble actions, high ideas, found worthy expression in strains of poetry and music, and the lives of illustrious men, the exploits of heroes, and exhortations to virtue, were written in verse and sung in public. Flute players accompanied the tragedies of Sophocles and Euripides, and the comedies of Aristophanes more than once borrowed the pith of their epigrams from the instruments that accompanied them.

As civilization advanced, human passions were gradually developed. New wants arose, and demanded hitherto unknown resources, the invention of musical instruments was then required, which would be more powerful and better adapted to give expression to various sentiments, higher thoughts, and a greater variety of ideas. The *tetrachord* formed, as its name indicates, of four tones or degrees, was the extent of the musical scale, and the only variety possible consisted in the order in which these four degrees followed each other. The three principal *modes*, (equivalent to what are called *keys* at the present day;) the Dorian, Phrygian, and Lydian, were distinguished by the place that the semitone occupied in the tetrachord; but being very limited and therefore incapable of expressing the passions and sentiments which, as has just been said, were constantly assuming new forms and different aspects, the original musical instrument was enlarged: higher and lower tones were added. The extent of the musical scale, which, as should here be mentioned, had never consisted of the natural octave, but of the tetrachord, was increased by a succession of tetrachords, a succession which imparted a peculiar character to the scales or keys.

The three original *modes* were extended. The "Dorian" was serious and grave, suitable for religious subjects, and according to Plato, who used it in his "Republic," adapted to the cultivation of good manners.

The "Phrygian," the character of which was ardent, proud, impulsive, vehement, and terrible, was used for military music, and found expression mostly in trumpets.

The "Lydian" was animated and piercing, but still rather inclined to softness. To these *modes* were added three others; the *Ionian, Æolian,* and *Mixolydian.* These new *modes* were in reality only mixtures of the three others, their founders; but they added to the extent of the harmonic system, they formed the basis of new combinations, and furnished artists with more powerful means for setting forth their ideas, and expressing their thoughts, before an audience, with more certainty and exactness. Later, several modes were again added: the Hyper-Dorian, Hyper-Lydian, Hyper-Phrygian, Hypo-Dorian, Hypo-Lydian, and Hypo-Phrygian.* But each one has characteristics peculiarly its own.

These various modifications, though they afforded increased resources to the art of music, were still far from satisfactory, for, as before suggested, with the advance of civilization came a corresponding growth of passions, and imperious wants, in the way of music, were felt, for sentiments and emotions which were no longer simple as before. A necessity was felt for finding some means of expressing on their musical instruments, these changes of emotion.

* As the Greek prefix indicates, these additional modes differ from the original ones by the position of the key-note; it being in one case above, in the other below. For instance —the key-note of the *Hyper*-Dorian is four tones above that of the Dorian, that of the *Hypo*-Dorian four tones below it. —TRANSL.

The tetrachord, although considerably augmented, failed to meet the demands made upon it; recourse was then had to invention. The *Psaltery*, an instrument of twelve strings, came into existence. This instrument, handed down from age to age, was almost exclusively employed by the ancients, to accompany the songs of the Jewish kings, and, after undergoing important modifications, assumed the form of our present harp. Music, among the Greeks, made constant progress, and, no doubt, would have attained equal excellence with the other arts, among this eminently artistic people, had there not occurred a revolution fatal to the country, and which arrested its onward course.

During a long series of ages, Greece—queen of the intellectual world, the sanctuary whence were evolved ideas destined to impregnate the genius of future generations, the school where countless philosophers came to study as well as teach, the cradle of eloquence and poetry, classical land of every moral and philosophical science, source of all human progress---Greece was to lose her intellectual supremacy, and to feel the sceptre which she had so firmly grasped, break in her hands. She was fated to a slow and sad decline. In the life of nations there seem to be bounds which cannot be passed. It is almost as if each nation had its allotted limit, beyond which it could not advance; as if, a certain end once reached, unknown to us, but still fixed and established, every people, every nation, was condemned, either by necessity, force, or fatality, to deteriorate physically, morally, and intellectually. History evidently proves it. Not only have these ancient races fallen into decay, but modern nations that have reached the very acme of glory and civilization, seem to-day to present signs of depression, lack of courage, dulness, and ignorance. Ah! may we never witness these sad phases in the history of our own country; may we never feel this approach of physical and moral degeneration! May we never see such fogs as marked the Middle Ages, shut down upon our own land, and our noble hopes sink into the abyss of fanaticism and ignorance! May we, in short, never lose sight of that blessed liberty, which too often is enjoyed but for a moment, and then vanishes like a brilliant mirage.

A city had risen up in the distance, menacing and terrible, of an encroaching and oppressive character. Rome, full of energy and vigor, consumed by ambition, greedy for conquest, had cast covetous looks toward the queen of the intellectual world. From this day it was all over with Greece.

The Romans, already powerful and urged on by an ungovernable desire for aggrandizement, aspiring to rule the world, rushed upon this beautiful country to make it subject to their own despotic sway. Little qualified to understand the eloquent beauty of the arts, completely ignorant of all that influences manners and civilization, those products of the intellect; having but a slight appreciation for the euphony of the language, summing up all their knowledge and aspirations in the following words: energy, courage, war, and victory, the Romans could not conceive of the artistic feelings of the Greeks.

But the daily association of the victors with the conquered, the beauty and mildness of the climate of the vanquished country, its monuments, masterpieces of grandeur and elegance, its national melodies, its admirable sculptures displayed on the pediments of the buildings and bearing the names of Praxiteles or Phidias, the dramatic works of Euripides, Sophocles, and Aristophanes, the thousand and one other productions of their genius, could not remain forever hidden treasures ; and at last, the Greeks and Romans began to understand each other.

Returning to their own country, the Romans, polished, though unconscious of it, by having been in contact with art, carried home the traditions which they had acquired. They were still ignorant how to put in practice what they had learned, but they possessed the germ destined soon to be developed. Meantime, the efforts of exiled Greek artists, or of those following the armies, were not without result. The Romans could not yet bring their minds to devote the time reserved for military exercises and manœuvres, to the cultivation of the fine arts in general and music in particular. Accustomed to military life, and skilled as soldiers, the rulers thought that raising armies, crushing weak and unoffending nations, invading near or distant countries, undertaking hazardous enterprises and adventures, killing, murdering, causing human hecatombs, and reducing races to slavery, were true grandeur. This fatal error, pervading even nations of the present day, made Rome the ruler and mistress of the world, the Rome which to-day is humiliated, debased, crushed by the most absurd despotism.

Nations, like individuals, need rest; like them, they are weakened and enervated by over-exertion; like them, they cannot be always in motion; therefore Rome, less disturbed by the fever of war, rested, or rather, grew meditative, and being in a state of repose gave itself up to reflection.

Its artistic feelings were aroused, and music, though it made no sensible progress, was cultivated at least to the same degree as when first introduced from Greece. In the forum, flute players accompanied the orators in their discourses, and the leaders of the people, through custom or vanity, were preceded or followed in their walks by bands of musicians. This movement ought to have been an indication of the progress and advancement of the art; nothing of the kind! There was nothing beyond a coarse and imperfect imitation, a pale reflection of the art as it existed in Greece. Foreign countries are not forcibly invaded, and their inhabitants deprived of their liberties, laws, and institutions, with impunity; no one can lay violent hands on their manners and customs, without incurring danger of retribution. Enslaved nations make up their minds to sigh and suffer in silence for a time, but sooner or later they become mutinous, make threats, rush to arms, utter a war cry, break their chains, and rise for freedom. It is not antiquity alone that has furnished examples of revolt and regeneration; our own age has given evidence of similar efforts and produced equally sublime scenes. Nations can for a longer or shorter time writhe in a slow and terrible agony, but they never die. Rome, powerful by its arms, almost always victorious, had incurred the hatred of the nations that she had enslaved but not subdued. The empire of force had triumphed; but however formidable this empire may be, it is never sufficiently so to be durable. Rome, execrated by the vanquished nations, roused all their anger and resentment, and became the one object which inspired their vengeance. Exposed to perpetual annoyance from people eager for her ruin, attacked on all sides by rude and barbarous hordes, seeking only plunder, overrun by masses collected from all parts of the world, the Roman empire began to decline. The name of Rome was not so widely extolled as formerly, the terror that it inspired was no longer strong enough to dismay her enemies even *before* a battle, and victory had more than once failed to crown her arms. In this sad condition of affairs, beneath the wave of barbarism constantly increasing and mounting higher and higher, music, which had never attained a very high excellence, even in the most glorious days of the empire, was utterly neglected in this age of distress and revolution.

There existed still some ancient traditions, and the harmony of the language, which had been improved, prevented indeed the entire disappearance of rhythm and cadence; but the incessant invasion of the country, and the voice and language of the invaders, had the effect of changing this delicious harmony day by day. The ear, constantly offended by discordant tones, lost, by degrees, all feeling of rhythm; and those that remained of the ancient Greeks and Romans became less and less sensitive to melodic accents. If any traditions were still preserved, they were powerless against the elements that tended to obliterate them.

Among the boldest and most intrepid invaders of the Roman empire, was a horde of barbarians called Gauls, consisting of a great number of independent tribes, but held together in several bands, all leagued in a kind of confederacy. This horde was aggressive, desperate and ferocious. Always ready to enter into contest, often victorious but sometimes vanquished, and then capable of avenging their reverses; whether triumphant or defeated, still making their enemies suffer, they left bloody footprints wherever they passed. Compelled for a time to yield to the power of Rome, they preserved their individuality, but had neither time nor mind for the cultivation of fine arts.

The life of excitement, the harsh and stern character, and the savage manners of these Gauls, could not be reconciled with the cultivation of an art which would be more likely to soften the manners than prove an incitement to war. And yet, if we are to believe their authors and historians, these proud and indomitable warriors had a music of their own. Bards, heroic and religious poets, ranking directly after the priests, had great and important influence, though to a less degree than the Druids. Guardians of the national traditions, their office was to celebrate noble actions and to apportion praise and blame. Their rapid verses, like the charge of their war-horses, as one poet expresses it, were sung in honor of their heroes. Their songs, which were accompanied by the guitar and harp, were capable of calming as well as exciting the most violent passions. "Often," said M. Henri Martin, "in the internal contests of the Gallic tribes, when with swords drawn, and lances lowered, the armies marched one against the other, the bards advanced between the two hostile lines, and quieted the fury of the soldiers, just as they would have appeased wild beasts by incantations." Of whatever effect it was capable, this music, the fruit merely of inspiration, could certainly not have any settled laws and rules, or any fixed principles. Their science consisted only of melody and rhythm.

Rome, constantly exposed to invasions and to incessant attacks from barbarous tribes, gradually wasted away: she lost her homogeneousness, as it were, and by constant intercourse with strangers her manners and customs became eventually modified.

In the midst of the confusion and intermingling of such various races, religion underwent considerable change. The polytheism of the pagans no longer satisfied their conscience, their temples were deserted, their religious ceremonies were no longer attractive. Olympus was passing away, or was nothing more than a dream of a few superstitious and ignorant fanatics.

A revolution, arising in secret silence, was to take place, and impart a new direction to human thought. It was an opportune moment; for the victors as well as the vanquished, exhausted by ceaseless strife, weakened by the loss of their choicest warriors, their intellect debased by incessant wars, clamored for rest and peace. All ardently aspired for the liberty and freedom which had been temporarily crushed beneath military despotism. A man of the people, born in poverty and lowly circumstances, is the first to utter the cry of freedom. He creates a new religion and preaches it. In the place of polytheism he teaches monotheism, liberty instead of slavery, equality in place of despotism, and brotherly love where tyranny formerly existed—all this on the basis of loving one's neighbor like one's self. Ascending from the creature to the Creator, he becomes absorbed in the contemplation of heaven, while he sunders the ties that bind him to earth. In his heart he seeks for inspiration, and finds there a temple and asylum. Principles, so different from those hitherto not only admitted but consecrated, moved all men who were brought to know them ; and therefore the new religion, resting on such a basis, worked a radical change.

It was opposed to too many interests, offended too many consciences, not to insure many decided adversaries, and even fierce opponents, and to call down much ill will and implacable hatred, so that for a long time it was kept down, repressed, and almost overborne. But there resulted from the persecutions an effect entirely opposite to that which was intended : many converts were made.

Rome, till then calm and indifferent to this change that was coming over religion, was greatly alarmed. Threatened in its religious faith, though no longer believing in it very sincerely, she endeavored to stifle, by the shedding of blood, the cry of freedom which had reached her very walls. She invented punishments, contrived tortures, made martyrs of the victims, whose cries were heard to the very ends of the world ; the blood poured out in the arenas only served to nourish the growth of the new-born idea.

In the midst of this convulsion of the pagan

world, music, confined to melodies handed down by tradition, was also in turn revolutionized. Christianity in its triumphant course had replaced all other forms of worship. Churches were erected everywhere. Cathedrals sprang out of the earth, as it were by magic; fervent and impassioned prayers filled all the temples; the most earnest vows ascended to heaven, and enthusiasm and love sprang up on every side.

For the purpose of giving thanks to God the Saviour and to Christ the Redeemer, music assumed a different character from heretofore. It had been materialistic, it became spiritual.

Education was not finished. Art having no rules, principles, or laws, advanced hesitatingly with trembling and uncertain steps, and all progress being stopped by old relics of paganism, there seemed no prospect of improvement in the future. However, about A.D. 384, St. Ambrose, bishop of Milan, while retaining all that was good in the old ideas, endeavored to introduce important changes in the art. Among the religious songs of polytheism, he chose the most popular melodies, and those most readily learned by ear and most fitted to the uncultivated voices of the multitude, and appropriated them to the worship of the new God; he adapted music to the words of the liturgy, and by this means the songs which had been learned and handed down by tradition were kept fresh in the memory. This process, still followed by the clergy of our own day, had a wonderful success. Unhappily, the holy father, while preserving the rhythmical music of the Greeks and the popular melodies of the ancients, while substituting sacred for profane words, restricted the art of music exceedingly, and condemned it to a dull routine.

Two hundred years after the death of St. Ambrose, another Pope, St. Gregory, endeavored to bring order out of chaos, and rearrange the songs and melodies in common use. St. Gregory, influenced by the idea of uniting the various Christian nations by means of the same language of harmony, invented a religious song, to which his name was given.* This song was divided into notes of equal value, and had no other rhythm

* Gregorian chant.

than that which inevitably accompanies human speech.

He collected a great number of melodies, making what he called an *Antiphonary*, and ordered that this collection should be considered a part of the liturgy. In consequence of this reform, religious music was the same throughout Christendom; it rested on foundations which were to all appearance solid, and yet the result was not what had been expected.

The *Antiphonary* did not fill every want; there was a necessity for having the musical language expressed by fixed and invariable signs. Numerous attempts were made, at different times, but all of them in vain, owing to the diversity of language, customs, and habits, among Christian nations. It was only toward the end of the tenth century that this great *desideratum* was apparently met.

Guido, a Benedictine monk born at Arezzo, according to the opinion of most historians, should have the credit of first finding a way to supply this long felt need about A.D. 1022. This was his manner of proceeding: he took the first syllables of the hymn of St. John the Baptist, and assigned a special and determined character to each. This hymn was composed as follows:

Ut* queant laxis
Resonare fibris
Mira gestorum
Famuli tuorum
Solve polluti
Labii rectum
Sanctæ Johannes.

It was learned by rote as a means of assisting the memory to recall and distinguish the sounds; the notes were named according to the syllables, and placed below them. The only aim of this kind of instruction was to impress the intonation of these sounds on the memory of the pupils. The names, *ut*, *re*, *mi*, *fa*, *sol*, *la*, were used to indicate the six notes of the scale; the seventh, *si* was only introduced into the musical scale, at a later period.

The importance of this innovation, or rather

* The syllable *ut* was exchanged for *do*, for the sake of euphony, by Gio. Maria Bononcini, father of the celebrated composer and rival of Handel.—TRANSL.

invention, is readily understood; it determined the intonation of every note, it enabled musicians, singers, composers, and instrumentalists of all countries to understand each other; it introduced one and the same language into the musical world, and therefore it was a radical revolution.

Materials were found; it only remained to construct the edifice. Each individual, each race, each nation, took hold of the work, and according to the taste, character, customs, passions, laws, and peculiar talent of each, musical composition, though limited to seven notes which had been settled upon and established, as has just been said, produced an infinite variety of combinations and effects. The art progressed and gave promise of a brilliant career.

As we pursue our research in the history of the art we shall see our way grow clearer.

Documents, though often unsatisfactory, still serve as *data*, and enable us to follow, step by step, the progress made by the human mind; but these brilliant gleams, torches as it were for us, are soon to be lost sight of, and disappear for a time. Social ties, which united the human race, are soon to be loosened, and a thick veil is to cover the greater part of our globe.

At the same time with Christianity, superstitious ideas were developed. The most grovelling fanaticism seized every soul, intolerance and error overshadowed all hearts; the despotism of the priests—chief feature of that age as of our own—knew no limits; society, shaken from its very foundations, disturbed by various religious ideas, bending beneath the oppressive yoke of absurd superstition and fatal prejudice, lost all power of reason, fell into ignorance, and gave itself up to the most extravagant speculations. Everything was dark and confused; and music, like all the other arts, sank into the common abyss.

The nations which had allowed themselves to be blindly governed by the church, and were living in a state of exaltation, through the influence of zealous priests, who believed themselves infallible, rose *en masse*, and rushed, like a torrent, toward the East. Under the pretext of rescuing the tomb of Christ, and delivering it out of the hands of the Infidels, thousands of fanatics engaged in bloody contests with the barbarous hordes, only to fall by the sword of the Saracen, or die under the sharp agony of famine, want, or pestilence. These great armies, constantly swallowed up only to be replaced by others, left their homes singing litanies and psalms; they returned, if it were granted them to return at all, singing *requiems* or a *de profundis*. What signified the dead, if the survivors believed themselves sanctified!

The art of music received new inspiration from these admirable, though fatal expeditions. Music assumed a new phase.

The Crusaders, fascinated by the graceful and ornate songs of the Orientals, began to imitate and repeat them; of course they did not fail to introduce them into their own country, upon their return. The *Trouvères*, *Troubadours*, and minstrels, like the Gallic bards of former days, sang songs of glory, tenderness, and love. The art became less severe in form.

The greater the distance from its point of departure, the colder became religious fervor, and the enthusiasm with regard to it subsided; man became more manly, so to speak. The age of blind faith was disappearing. The accumulated dust of ages, which concealed all traces of civilization, was blown away by the powerful breath of a social condition which was undergoing regeneration, or rather transformation. The fear and terror inspired by the tribunals of the Holy Inquisition had no longer their former power; the flames of funeral piles no longer shone with the same brilliancy; the pincers and instruments of torture began to rust in the sacred armories, and the cries of the victims were only heard at long intervals. A movement of regeneration had been commenced with no common energy.

This regeneration was not confined to what concerns material and philosophical life. It was manifested, as well, in all that pertains to the moral and the intellectual.

The masterpieces of antiquity, abandoned and forgotten for a long series of years, were still in existence. They afforded material and served as guides for the erection of new edifices. The artists were the first to brush the dust off the monuments, and sculpture, architecture, painting, poetry, and music, became objects of study and close imitation.

The imitation was not, nor ought it to have been, servile, for the monuments and intellectual productions of the ancients had neither the same end in view, nor the same source of inspiration.

The ancients, impassioned interpreters of physical beauty, were skilful enough to reproduce it to the greatest perfection, the artists living at the time of the *regeneration* just mentioned, inspired by the beauty in nature, as well as Christian sentiment, attempted to express, in their works, nature idealized.

As to the philosophers of the same period, while they admired the works of past ages, they were no longer humbly submitting to the sway of Plato and Aristotle; they accepted, it is true, their principles and doctrines, but not to such an extent as to be trammelled by them.

This earnest movement of emancipation did not affect men alone; the women, who comprise what is called, according to a great and celebrated physiologist, "the sensitive portion of humanity," were not behind. They plunged into the study of arts, letters, and sciences, with all the ardor and passion that always mark their actions. The courtesans were the first and the most eager to seek for new and powerful means of fascination, in the cultivation of poetry and music.

In the arts, there was a tendency to replace with the ideal, that strict adherence to the materialistic, so to speak, which is manifest even to-day in the pediments of the basilicas, in the capitals of the columns, or in the arches of the Gothic cathedrals. In the sciences as well as in philosophy, a spirit of analysis was brought to bear on all research; which was aided by earnest reasoning or argument, as well as a strict devotion to nature.

The old feudal system was gradually disappearing, under the powerful influence of this regeneration; ideas became more clear, customs less austere, and bright-colored garments, of elegant shape, replaced those of a sombre hue.

Social gatherings increased, dramatic performances were of a more pleasing character, rich and elegant palaces arose on all sides, and poetry and dancing added new attractions to these splendors.

Light appeared, darkness and obscurity vanished, a more humane feeling took the place of an exclusively religious sentiment, and two powerful tendencies were manifested in a most active and energetic manner; the aim of both being the emancipation of the mind enslaved by the spirit of fanaticism, and the enfranchisement of man oppressed by despotism and the priests.

Faith was no longer strong enough to silence reason. The free exercise of intellectual liberty was almost the sole aspiration of European nations. The chains of absolute power, too heavy to be borne, were on the point of being broken. Sweeter and more worldly pleasures followed the asceticism of the Middle Ages, which encouraged mortification of the flesh and spirit. The accents of prayer were almost drowned by songs of gayety; and music, carried along in the general movement, gave expression to the common frivolity and, one might almost say, libertinism.

This music was heard, not only in gay and lively society, but it pervaded even the churches. The infatuation was such that music possessed but one characteristic—that of levity. It was even applied to sacred words, of which it changed the sense, and imparted to them a licentious meaning; sometimes even whole masses were written full of obscene expressions. In a *Sanctus*, or *Incarnatus est*, words were introduced which decency or propriety would forbid our repeating here.

Things finally reached such a scandalous and vile state, that the Councils, especially that of Trent, launched anathemas on the abuse of music in the churches. They branded with a special stigma of disapproval the low custom of mingling profane words and music with sacred text and song.

Pope Pius IV., obeying the express wish of the Councils, appointed a commission to devise means to be employed for the reformation of these abuses, and he appealed to a musician for aid, who had been long known through his musical works, which bore the stamp of the most fervent religious feeling. Palestrina, recognized and appreciated by all the world on account of the superiority and grandeur of his sacred songs, had attracted the attention and gained the patronage of Pope Marcellus, predecessor of Pius IV. He was commissioned to restore the spirit of devotion and reverence which had long been missing from church music. He immediately set about the work, and composed three masses full of charming melody, and deeply religious in character. The third of these masses, generally known as the *Mass of Pope Marcellus*, thus named by its author out of regard for the memory of the pontiff, was considered the best, and was therefore adopted as a standard of excellence.

At the first hearing of this composition, June 15, 1565, Pius IV. was so enraptured that he summoned the author of this masterpiece before him, complimented him, and appointed him director of music in the Papal chapel. At this time Palestrina could not regard such an honor with indifference. Happy in the success he had just attained, he pursued his artistic work with the greatest ardor, and his efforts all bore the stamp of genius.

The fifteenth and sixteenth centuries are considered, in the annals of the human race, a brilliant and remarkable period with regard to intellectual productions. The fine arts in general, and music in particular, advanced more rapidly than ever before. Great events and new discoveries helped to overturn and transform the whole world of intellect.

In less than half a century, printing, invented in 1450, offered a means for a tangible and palpable manifestation of human thought. It put all nations in communication with each other; it brought them nearer together, and allowed them to participate mutually in the progress of human science; it caused another complete revolution.

Three years later, in 1453, the capture of Constantinople by the Turks revealed to Europe Grecian civilization, and disclosed models full of grace and instruction; it also shed light on the philosophy of past ages, brightened the minds of artists, and inspired a taste for study and research.

In 1492, the discovery of America, which revealed, as it were, a new world, opened a vast field for exploration. A virgin land, free from all stain, a splendid and luxuriant vegetation, a state of wild and vigorous nature, presented an infinite future to the imagination, and a never-failing spring of varied inspiration to the mind.

The movement of immigration to this wonderful country, though impeded at first, soon set in with an almost passionate energy. Unfortunately, the race who were the first to tread these heaven-blessed shores, possessed neither the germs of poetry nor the ideal of thought; entirely occupied by commerce and trade, they saw only another means for obtaining material wealth in this newly discovered country, only another outlet for their productions, only another market to be supplied. America for a long time was only a vast factory, and, even in our own day, is merely an extensive work-shop, admirable, it is true, and whence issue the most astonishing and marvellous products of human industry.

Europe, twenty or thirty years later, underwent a startling change of faith. A bold mind dared to question the truth of Catholicism, which had hitherto held triumphant sway. Luther presented his views of reform to the religious world. It is true, they bore only on points which admitted of dispute; but they imparted a shock to the Catholic world, which is felt even to day. It could not be otherwise; for the doctrines and dogmas of this church, once submitted to reason, analysis, discussion, and open attack, Catholicism ceases to be a revealed religion, and becomes simply a philosophy, whose teachings admit of every-one receiving instruction in the supposed truths, or bringing the light of his own opinions to bear upon them. Therefore the spectacle which we witness to-day, in our own age, of science carrying with perfect right the world-illumining torch, ought neither to irritate nor astonish us. The *non possumus* of the sovereign Pontiff is the inevitable consequence and fixed principle of the Catholic religion; that is, an absolute denial of science—this illuminating torch—as well as the right of discussion, which guides and directs it.

The Reformation, attacking the faith which accepts without question all that it is told to accept, drew upon itself the most violent hatred and the liveliest animosity. The strife was long and bitter, persecution was incessant, tortures were renewed, and there was no lack of victims.

The movement, which affected all humanity, opened up a new life to society in general; philosophy resumed arguments long forgotten; art began again to advance, after having been for a time interrupted; science took a fresh start, never again to be arrested in its onward march; the spirit of independence pervaded the masses; the principles of liberty were engendered in all hearts; in short, the hour for a new birth, a complete regeneration, had struck.

These events, and many others that could easily be recounted, were we to penetrate the recesses of this remarkable age, would seem perhaps to have no connection with each other; yet they were signs of the beginning of a long needed investigation; they were all so many symptoms of hope, or aspirations which could no longer be kept at rest by an authority always considered until then absolute and all-powerful.

Music, like all the other arts, became emancipated, as it were; and the great reformer, Luther, convinced of the powerful influence of an art, of which he himself was passionately fond, began composing chorals for the churches. Under his inspiration, schools of music were established, bells and chimes were placed in all towers and belfries. The Catholic States of Germany, not willing to be outdone, followed the example of the Protestants, and instruction in the musical art formed, with them too, a part of common education.

Music, which had long been stationary in Germany, was destined soon to take a leading position. The sentimental character of the Germans predisposed them to instrumental music, and they had a special love for symphonies. Among this people, in fact, the dreamy, poetic, and ideal had full scope for development. Bach,

Handel, and Haydn, were worthy and illustrious representatives.

Profane music, expelled from Christian churches, and prohibited by order of popes and councils, as has been said above, took refuge in private houses and social gatherings. There, being favored and upheld by the fashion of the day, it became established on a firm basis. Musicians of natural taste, who were led to study and cultivate with earnestness an art that afforded them both enjoyment and pleasure, constantly invented new musical combinations.

The epoch of blind faith and belief had already gone by, and the mind was no longer absorbed by religious emotions. Religious wars gradually ceased, there were no longer continued arguments among scholars, civilization had resumed its march, new passions had come into play, wants were increasing, more intense longings and aspirations were felt; as a consequence, more variety and energy were demanded in musical tones and accents. It was expected to interpret all sentiments or emotions, to express every situation in life, to adapt itself to all characters, and to identify itself with whatever subject was to be considered or represented.

Greater demands were made upon both artists and art itself, for the purpose of imparting more enjoyment to the world at large. Musicians, on their part, understanding better what was expected of them, multiplied their resources. They no longer confined themselves to giving expression to their ideas and thoughts by sounds alone.

It was especially in Italy that the most striking advancement was made.

Monteverde, a man of immense talent, it might almost be said of genius, born at Cremona, 1505, a skilful *virtuoso* on the viola, was summoned to Venice, and appointed director of music in the Cathedral of St. Mark, August 19, 1613. He was one of the first to impart to music a much more significant meaning than it had ever before had. Beside composing in all the styles known in his day, he attempted writing works in dramatic form.

At the court of Mantua, in 1607, he brought

out an opera, *Ariadne.* Soon after, *Orpheus* appeared. The greatest success crowned these efforts. In 1608, on the occasion of the marriage of Francis of Gonzaga and Margaret of Savoy, the same artist composed music for a *ballet* called *Delle ingrate,* in which, according to the opinion of competent judges, effects of rhythmical instrumentation were produced which had been totally unknown up to that age. This *ballet* was received with enthusiasm, and warmly applauded.

The impetus was given; musicians, animated by an earnest and noble zeal, worked most industriously. Each chose his own path; some had more taste for religious subjects, others for heroic ones; some for pictures of love, others again for pastoral science, and the aim of all was glory and renown.

Dramatic music had been invented; the drama, varied in form and character, was destined to have countless interpreters.

Toward the end of the sixteenth century, Don Garin de Toledo, viceroy of Sicily, had the *Aminta* of Torquato Tasso, and a *Pastorale* of

Transillo, brought out in presence of the court. The Grand Dukes of Tuscany, considering themselves protectors of the arts, in imitation of Mæcenas, patronized all attempts in the way of dramatic music most generously. The art of musical composition was developed and advanced on every hand. Harmony became more incorporated with melody, and thus accompaniments were composed that presented marvellous effects. Instruments were subjected to pleasing modifications. Their compass became more extended, their tone more ample and rich. The nations who witnessed the success of these efforts commended the diverse changes; they followed the progress of an art, that had already afforded them so much pleasure, with the keenest interest. Italy was the first to distinguish itself, and the Neapolitan school had the honor of being at the head of this progressive movement.

Every day witnessed the appearance of new operas, oratorios, and motets. Compositions were written in all styles. The number of illustrious musicians soon became so great as to preclude the possibility of their being either named or classified.

While Naples was advancing with rapid strides, and the art which had made its school renowned was being cultivated with eminent success, Venice saw numerous aspirants rise up and become famous, in its very midst. Benedetto Marcello, a nobleman sprung from a partrician family that included among its ancestors a doge, six procurators, and a number of other illustrious civil and military dignitaries, gave promise of being a distinguished composer. Carried away with enthusiasm for an art which he loved almost passionately, he had to contend against the prejudices of his time and the obstinate will of his noble father. The nobility, being proud and haughty, thought it in those days derogatory to their character to engage in anything but the pursuit of pleasure; they considered it degrading to take part in the cultivation of an art, in which even members of the lower classes often distinguished themselves.

Benedetto Marcello obstinately endeavored to

break down the prejudices of his family, and remove the obstacles that stood in his way, but seemed powerless to overcome the objections of his father. However, such an indomitable desire to express the musical ideas which seemed pent-up within him, must necessarily triumph in the end. He secretly bought some music-paper, and at night devoted himself to his favorite pursuit. His father, discovering that he was disobeyed, removed all means of work from his rebellious son; he took him away some distance from the town, hoping to divert him, and distract his attention from studies which were unworthy, in the father's opinion, of his son's high birth. Useless measures of precaution and severity! Benedetto, more inspired than ever, when living in close contact with quiet *nature*, contemplating the calm peacefulness of the meadows, felt a more intense ardor burning within him. Ideas and melodies were constantly crowding upon him, and for the second time, he secretly procured some music-paper. Far from the excitement and noise of the town, in reverie and solitude, he wrote compositions which were destined to be master-works. In the midst of this ceaseless opposition; these obstacles which would have been insurmountable to any less determined will than his own, and this continuous vexation, young Marcello was called upon to witness the death of his father. Then the unfettered genius took wings, and a mass of his, which was highly applauded and regarded as a masterpiece, revealed the immense talent of this nobleman-composer; the success he met with, the applause, and later, the renown and glory that were awarded him, recompensed the artist for all his past sorrows. Benedetto Marcello, free now to devote himself exclusively to the profession for which he had shown such genius, became wholly absorbed in this one passion.

A thorough and varied education, acquired by long literary application, enabled him to understand intimately a great variety of subjects. Masses and oratorios opened up a vast field to his inspirations, and the music he adapted to the Psalms of David gave evidence of a brilliant imagination, a soul full of noble and high thought, a great warmth of heart; as well as a character remarkable for energy, tenderness, and sensitiveness. This man of such wonderful powers died in 1730, at the age of forty-four, with his mental faculties undiminished.

Italy shone so brightly that its lustre was shed over all Europe. France and Germany, at first cold and indifferent spectators of this great movement in musical art, eventually took part in it.

Cardinal Mazarin, minister to Louis XIV., was the first to introduce dramatic music into France. He looked toward Italy, summoned singers and musicians, and had the dramas already celebrated in the Peninsula performed in the presence of the monarch so long oppressed with *ennui*.

The *Festa teatrale della finta pazza de Strozzi*, and the *Orpheus* of Monteverde, inaugurated an epoch which was destined to be fruitful in musical compositions. A lyric tragedy was played at the Louvre, at the time of the king's marriage ; pastorals appeared under the inspiration of skilled musicians and the all-influential patronage of the prime minister.

From this day dramatic music received the right of citizenship as it were, in France, or rather, the privilege of being identified with court circles.

While these attempts at representing and interpreting the lyric drama were becoming more frequent, and meeting with eminent success, a distinguished violinist, Florentine by birth, was brought to Paris by the Chevalier de Guise. His reputation as an artist gifted with delicate sensibility and perfect taste, had preceded him; therefore Lully could not fail of success. Courted on account of his wit, as well as his great learning in musical matters, he was soon appointed Royal Director of Music. A short time after, he is announced as filling the office of court composer, though nothing was actually required of him. Finding himself all at once in the midst of a court devoted to pleasure, he wrote accompaniments to the ballets given there, as well as arias to be introduced at the representations of Molière's comedies. A man of vast resources, and gifted with great powers of invention as regards melody and harmony, he was granted the privilege of making use of the Royal Academy of Music, where he brought out, first, a pastoral by Quinault, called *The Feast of Love and Bacchus.* The first performances met with most flattering success, and the poet and musician, working hand in hand from this time, as is the case in our own day with Scribe and Auber, gave to the world operas, pastorals, and no less than twenty-five ballets, in quick succession.

Singers and choristers, dancers and orchestra players, had each their special part to perform. Opera was now fully established.

Dramatic music became the favorite among all classes. Frightful plots, pastoral scenes where love reigned, fascinating ballets, and charming interludes, followed one another in quick succession on the stage. Lully was unquestionably the all-ruling spirit.

But in the midst of the success and triumph of this Florentine, worthy representative and interpreter of the Italian school of music, a French musician suddenly came into renown, owing to his theoretical works on the art of musical combinations. Harmony was the special object of his study, but the obstacles which constantly arose in his path prevented his putting in practice his theories and principles. It was not until Rameau was forty-four years old according to some writers, but fifty according to others, that he brought out an opera at the Royal Academy of Music. His first work was presented to the public in 1733, and it was followed by twenty-four others, all of which bore the stamp of a high order of genius.

Rameau, more intense and passionate than Lully, and imbued with the grand and eternal principle proclaimed long before by Monteverde and his followers, that music should only obey feeling and have no office beyond that of accompanying poetry and rendering its meaning more clear, had an advantage over Lully, in being able to invest his heroes with more force, and impart more passion to his melody, while his harmonies were more rich and complete.

These qualities might well be expected to attract the attention of *connaisseurs* to his works, and to win the admiration of all hearers. The public became wildly enthusiastic, and, as always happens under similar circumstances, a powerful reaction soon took place.

Lully's works, lately admired to the very exclusion of all others, were quietly laid away on the shelves of the Opera House, and their master was almost entirely neglected.

Rameau's music enjoyed an incredible popularity for some time, and reigned without a rival until a troupe of Italian singers obtained permission to give dramatic performances on alternate nights with the French company at the Opera House. Lully's works were again brought from the oblivion into which they had been unjustly cast, and appeared once more on the scene. Other more modern Italian compositions aroused a little of the former enthusiasm in the public. The *Serva Padrona*, undoubtedly the master-work of Pergolesi, was brought out by the Italian troupe August 2, 1752. It was listened to with unbounded satisfaction and delight. Other works by the same master were also given, and from that day French music was to hold an equal rank with the Italian.

Two parties were formed. The French party bore the name of the "King's Corner," and the Italian one that of "Queen's Corner," because each took its place respectively near the King's and Queen's boxes at the opera house. At this epoch, Madame du Barry, who exercised a strong personal influence over the king, pronounced in favor of French music, not that she in the least fancied Rameau, on the contrary she detested his works, but out of politeness to her royal lover.

The two parties were constantly stirring and active; pamphlets and all sorts of satires appeared in every direction. The Italian music was attacked as being too light and superficial, partaking too much of the character of the *opera bouffe* of the present day; French music was considered as tiresome, heavy, formal—too much resembling psalmody. Critics, musicians, and the most renowned literary men of those times, were all interested in this quarrel, as it might be called; their opinions, were openly given, and their wits expended and pens freely used for the benefit of both parties. J. J. Rousseau, who belonged to the

"Queen's Corner" wrote letters on music. Grimm and Voltaire, who were always to be seen in the "King's Corner," were constantly launching various articles and satires against Italian music. *Ramists* and *Lullists* entered upon really a fierce war; each party defended its own views with an almost unparalleled pertinacity. Voltaire, tired of such constant quarrelling, tried in vain to bring about a reconciliation; he wrote at one time:

"Je vais chercher la paix au temple des chansons,
J'entends crier; Lully, Campra, Rameau, buffons,
Etes-vous pour la France ou bien pour l'Italie!
Je suis pour mon plaisir, messieurs, quelle folie
Vous tient ici debut, sans vouloir m'écouter?
Ne suis-je a l'Opéra que pour y disputer?"

But his words were lost in the tumult that reigned. The contest continued to rage with the greatest violence and ardor, and at the time of the production of *Titon and Aurora*, January 9, 1755, the quarrel fairly reached its height.

The first representation of this opera was considered as decisive either for triumph or defeat.

Both sides prepared themselves for the struggle. When the day came for the performance, the two parties were in their accustomed " corners."

" The light-horse," said M. Castil Blaze, " the gendarmes, musketeers, and riflemen filled the parquet." The Italian, or Queen's party," crowded out of their usual " corner," and literally unable to find a place anywhere in the hall, took up a position in the corridor. Excitement and impatience had reached the climax. The curtain rose at last, and *Titon and Aurora*, thanks to cunningly devised tricks, which were made use of by force, proved a success. During the course of the performance, the *dilettante* portion of the French party despatched couriers every fifteen minutes to Louis XV., who was at Choisy, to advise him how matters were going on.

The monarch, impatient to know the result of the contest, in which he had taken such a keen interest, awaited each courier as anxiously as if some serious state-question were under consideration, or the very fate of his kingdom were at stake. " We have gained our cause," he cried, addressing his favorite, after having received the last bulletin announcing the final triumph.

Considering all the artifice used to accomplish this success, it could hardly be called a brilliant one. However, it effected an immediate result: the complete dismissal of the Italians. French opera was granted the exclusive privilege of the Parisian stage.

This victory, though won rather ignobly, was productive of new favors for Rameau: the king presented him with a patent of nobility, a necessary preliminary to render him worthy of receiving the Order of St. Michael which was to be offered to him. But the musician, not sufficiently appreciative of these marks of regard and esteem, did not even have his patent of nobility recorded. Louis XV., thinking without doubt that Rameau did not wish to incur the necessary expense, proposed himself to defray it. "If your majesty," said the artist, "would really like to present me with this sum of money, I could spend it much more advantageously. What is a noble *title* to me? *Castor* and *Dardanus* have long since added sufficient lustre to my name." However, the Order of St. Michael was presented, though the title of nobility lapsed, owing to its never having been recorded.

This artistic pride, and this indifference amounting almost to contempt for ribbons and decorations, were by no means rare at this period. Tintoret refused an order which was offered to him. Later, the sly and witty Beaumarchais, whom a certain critic had described as not gifted with a bright mind, and as being a theorist of narrow views, replied to Baron de Breteuil, when that minister offered to present him with the Order of St. Michael, "Those men whose merit is *recognized*, can well afford to dispense with decorations, so precious to fools" (June, 1785).

Have artists of the present day any such scruples? Let each reply for himself. *Tempora mutantur, et nos mutamur in illis.*

After such long quarrelling, by means of both voice and pen, the point of the satires seemed to be blunted with regard to Italian and French music, and both *Ramists* and *Lullists* appeared to have tacitly consented to a truce, when a new occasion came to revive the quarrel.

A composer arrived in Paris who was well-known in Germany and Italy, where he had already brought out several works with great *éclat*.

A powerful reformer, he left the worn and beaten path hitherto travelled by musicians, and resolutely broke a new one for himself.

Combining originality with genius, heralded by his works, Gluck was about to appear before the Parisian public, and to undergo the criticism of the two parties, between which ill-feeling had been allayed but not extinguished. Therefore curiosity was on the *qui vive;* every one was prepared for attack or defence.

The opera of *Imphigenia in Aulis* was announced for April 19, 1774. As early as five o'clock the entire court were in their respective places. There was a perfect blaze of gold, precious stones and diamonds. The *ban, arrière ban* and the *dilettanti* met together ; friends and foes alike were at their post. The buzz of conversation was heard through the house, but at a signal given from the stage, all became quiet, and a solemn silence succeeded the confusion of tongues. The overture began, and the opening chords filled the audience with admiration.

The hurried and excited breathing of the listeners was apparent, but no one dared give expression to his feelings. After the first recitative of Agamemnon, the wife of the Dauphin, Marie Antoinette, fairly carried away with enthusiasm, clapped her hands. Then the excitement knew no bounds; a thunder of applause drowned the voices of the singers and the notes of the orchestra; the success was decided; *Iphigenia* was cheered; Gluck triumphed!

This performance, which the *Annals of the French Opera* mention in the most enthusiastic terms, justified the opinion first formed of the genius of this composer.

Pure delivery in the recitatives, power in the dramatic expression, wealth and inspiration of melody, exquisite refinement of sentiment and emotion, all contributed to the well-merited and enthusiastic celebrity of the artist.

Louis XV. still lived; Madame du Barry, having witnessed the enthusiasm and applause that greeted the protégé of Marie Antoinette, for whom she entertained the most violent antipathy, wished, out of a sheer spirit of opposition, to introduce a musician who should be exclusively under her own protection. In connection with the Marquis Caraccioli, the ambassador from Naples to Paris, she had a letter sent to the Baron Breteuil, French ambassador at the Court of Naples, telling him to enter into negociations with Piccini, already known in Italy through his remarkable compositions, and highly praised by the banker Delaborde; an annual salary of two thousand crowns, and the privilege of having his works copyrighted, were offered the maestro, if he would bring out his operas at Paris. Piccini accepted this advantageous proposal, and thus the Italian and French schools were again brought into opposition with each other; or rather Piccini was pitted against Gluck.

The rivalries which might be expected to arise between these two champions of the different schools could not fail to be exciting, and inspire those lookers-on who took an interest in the success or defeat of either party, with the keenest emotion.

The arguments which had been already advanced for and against the French school, were all to be brought forward in connection with Gluck's music, and the complete and brilliant success he had just obtained did not serve to discourage the partisans of the Italian school. They contended that the great composer's music was too severe in style, and clamored loudly for more lightness, mirth and airiness; so they had great faith that the music of Piccini would answer all their demands.

Piccini, a musician of almost exhaustless resources, gifted with a brilliant imagination and, above all, a wonderful originality, immediately set to work, and soon after arriving in France, brought out the opera, *Atys*.

With the announcement of this coming novelty

the court and citizens assembled at the theatre; the music of the new master was greeted with the greatest applause. This was possibly all the more noisy, because the enthusiasts aimed to have the reception of this new master outshine in every way that tendered recently to Gluck.

Musical quarrels now became more animated than ever. *Gluckists* and *Piccinists* entered into angry dispute. Like the *Ramists* and *Lullists* they had among their friends and adversaries, authorities as regards literature and criticism; on both sides, anger was violent, fierce, spirited and passionate. If we wish to form the least idea of the fury of the contest which then arose, we should revert to the days of our youth, when we witnessed the quarrel between the *romanticists* and *classicists* in our French literature.

Marmontel, that zealous advocate of the Italian school, thought he could take the best operas of Quinault and adapt them to the style of Piccini; this called down upon him the animosity of the critics who were partisans of Gluck. Arnaud, in one of his satires, wrote:

"Ce Marmontel, si long, si lent, si lourd,
Qui ne parle jamais mais qui beugle,
Juge la peinture en aveugle
Et la musique comme sourd
Ce pedant à si triste mine,
Et de ridicule bardé,
Dit qu'il a le secret des beaux vers de Racine
Jamais secret ne fut si bien gardé."

The struggle was carried on in every possible way. The Italian party was eventually vanquished. Gluck, the illustrious composer of *Armida*, *Alceste*, *Iphigenia in Aulis*, and *Iphigenia in Tauris*, was universally applauded.

This ardent and passionate strife did not fail to have some effect on the future progress of music.

An extensive field was opened up to talent and ambition of all kinds. Men of undoubted genius entered the arena striving for fame, and in that age, as to-day, success in opera was the aspiration of all musicians—the goal, as it were, of all competitors.

One of them, a young man hitherto almost unknown, was preparing silently and in secret to enter the lists. A passionate admirer of the works of Gluck, he eagerly improved every opportunity for listening to the compositions of the great master, whom he regarded as a model.

An anecdote acquaints us with the admiration which the young man entertained for the musician. whom he worshipped almost to idolatry. The first representation of the opera, *Iphigenia in Tauris*, was announced. Our would-be musician was seized with an irresistible desire to witness this performance. Unfortunately an insurmountable obstacle prevented him from satisfying his ardent curiosity. He had no money with which to buy a ticket. What was to be done? Make use of stratagem. The day before the performance, the young man secreted himself in the furthest corner of an opera box. What matters it to him, if he must pass the night and day there without food or drink, provided he can witness the play without any expense? While he is crouching in his place of concealment, a noise is heard; it is the inspector of the theatre, making his rounds. The culprit's heart beats faster: he is discovered! The inspector takes him to Gardel, the director of the

Opera. "What were you doing there," asked the latter. The young man, blushing with shame, replied that his pecuniary means not allowing him to purchase a ticket, he concealed himself in the box so as to witness the play. Gardel, with his usual magnanimity, presented him with a ticket, which was eagerly accepted. The name of this young, until then unknown musician was —*Méhul.*

Being presented to Gluck, he was kindly received, and was even granted advice with regard to the study of his beloved art.

Being gifted with original talent, and faithfully following the counsel and instruction of his master, Méhul learned how to give expression to the highest kind of beauty. "Dramatic sentiment," according to a biographer, "vigor, truthfulness of expression, imagination, boldness in orchestration, the wonderful effect consequent upon the skilful combination of the three powerful elements in music—melody, harmony, and rhythm—all are to be found in his compositions. Since his time, our means of working have become more numerous, powerful, and varied, but we have never overstepped the limits fixed by Méhul. His opera of *Euphrosine* is full of purity, grace, and excellence" (Castil-Blaze).

Méhul, whose talent and merit have been appreciated by all the world, was the first to introduce overtures as preludes to operas. Nearly all of these overtures are masterpieces, and we still listen to those preceding the operas of *Stratonice*, *Horatius Cocles*, *Young Henry*, and *Joseph*, with the greatest delight.

Méhul did not confine himself to operatic compositions. Carried away by the revolutionary movement, it filled him with thoughts and inspiration, the burden of which was earnest patriotism. Among his numerous works, that which met with the greatest favor from the public was, without doubt, the charming hymn, constantly heard wherever any enthusiasm or passion is felt; that known as *Le Chant du Départ*.

Music, as we see, assumed new forms day by day; the creative zeal of artists and the admiration of *dilettanti* was everywhere aroused; the

Royal Academy of Music was no longer able to bring out all the new works and dramatic compositions that were offered. It was powerless to satisfy the ardent curiosity of a public, now so devotedly fond of an art that afforded them such exhaustless enjoyment. Some other means had to be provided for representing the countless productions which were daily brought forth, and therefore a separate house was established for Opéra Comique. All the works were there put upon the stage which did not properly belong to the Grand Opera. One of the first musicians who met with the greatest success on this newly created stage was undoubtedly the *maestro* Grétry. He had written many compositions for the Grand Opera, but they had never received the applause which their undoubted merit deserved. The operas of *Andromache*, of *Cephise*, and *Procris*, spite of the learning of which they gave evidence, were coldly received, and *L'embarras des richesses*, written in conjunction with Marmontel, called forth the following epigram on its authors :

"Embarras d' intêrêt,
Embarras dans les rôles;
Embarras de ballet,
Embarras de paroles,
Des embarras, de sorte
Que tout est embarras;
Mais venez à la porte
Vous n'en trouverez pas."

Collinette à la Cour was not as successful as the other pieces, but the *Caravane du Caire,* produced at the Opera Comique January 15, 1784, met with the greatest favor, and was loudly applauded. This melodious, lively, graceful and brilliant music, was particularly remarkable for its truthfulness of expression.

Monsigny, Philidor and Berton, that constellation of brilliant musicians, to speak poetically, inspired the masses with a taste for this kind of music, which, in our day, still touches the sympathies of the majority of music lovers.

In Italy nothing was heard in the theatres but the compositions of Guglielmi, Paesiello, and Cimarosa; which works, full of true inspiration, grace and originality, were everywhere admired.

After Gluck's departure from France, Piccini reigned triumphant. His *Dido* placed him unquestionably far above all cotemporaneous artists, but Sacchini was now to contest with him for the leadership of the opera at Paris. This composer, one of the most renowned musicians of the age, brought out operas which were crowned with the most brilliant success. He was a distinguished pupil of the Conservatory of Music at Naples, and had travelled through Italy, Germany and Holland; the report of the success and applause which he had met with in these different countries, had even reached England, where dramatic music was just beginning to be appreciated.

During his residence there he composed a large number of *serious* operas, and a few works in *opera bouffe;* among the first, we may mention as the most renowned, *Montezuma* and *Rinaldo;* among the second, *L'Amore soldato* and *La Contadina in corte.*

Leaving London after a sojourn there of eleven years, he went to Paris, where his well-deserved renown was not unknown, and brought out, on the

French stage, works that charmed and delighted the public by richness and elegance of orchestration, as well as grandeur, purity, and sweetness of melody. The name of Sacchini became popular, and his operas, *Chimène*, *Renaud*, and particularly *Dardanus*, a subject which had been previously treated by Rameau, sealed the reputation of the celebrated composer. But, alas! jealousy arose simultaneously with his popularity; outcries by interested parties, under the pretence of protecting the honor of national artists, and advancing their prosperity, prevented the representation of Sacchini's operas at the Royal Academy. Grief fairly overcame the master, and aggravated the fever from which he was suffering at the time, so that before he was allowed to witness the performance of his masterpiece, *Œdipus at Colonna*, Sacchini was gathered to his fathers.

In our country, which is considered moderately intelligent, ought the spirit of envy, detraction and injustice to assail all who give any evidence of genius, or already triumph in the consciousness of fame?

At this time Salieri, the director of the theatre at Vienna, in Austria, was acknowledged by foreigners to be a distinguished composer, but his excellence had not yet been as widely recognized as it deserved.

A friend of Gluck, and in constant association with him, Salieri composed, in 1784, the opera of *Les Danäides*, for the Royal Academy. The music in it was so remarkable, so powerful and so full of beauty, that it was attributed to Gluck, until the latter solemnly declared that the work contained nothing but what was wholly original with the musician who claimed its authorship.

Three years after this great success (1787), Salieri wrote an opera called *Tarare*, the libretto of which was from the pen of Beaumarchais; this opera was eminently successful, though to a somewhat less degree than *Les Danäides*.

Something now occurred which, according to Castil-Blaze, had never been known in the history of art. Beaumarchais and Da Ponte, separated from each other by four hundred leagues, hit upon a style of operatic composition called the *romantic*.

Beaumarchais and Salieri had just produced *Tarare* June 8, 1787, when the sublime *Don Juan*, joint work of Da Ponte and Mozart, appeared upon the stage at Prague, November 4, of the same year.

While the musicians mentioned above were at the height of their glory, Europe resounded with the fame of a young man who, even at five years of age, was a distinguished pianist, and two years later, a composer. Mozart was now in Paris, for the second or third time. Introduced to Grimm, and to Noverre, *maître-de-ballet* at the Royal Academy, as well as to Madame d'Epinay and several other distinguished persons, this young musician begged the favor of being allowed to execute some of his compositions in their presence.

For this purpose he placed in the hands of Legros, director of the classical concerts, a symphony written for the most eminent instrumentalists of the age. Predictions were freely expressed on the part of Mozart's patrons ; every-one seemed to be favorably inclined toward him ; they even talked already of permanent engagements ; but all these good words were, so to speak, empty promises. Time passed, no engagement was formed, the hour for fulfilment never arrived. What seemed still more strange and provoking, was that they did not even deign to have the score of the symphony copied, though it had been received by the public with something like enthusiasm. Mozart, disheartened by six months' delay and vexation, had no further strength to plead his cause. He could not even endure remaining longer in Paris. His trials and disappointments, coupled with the death of his mother, with whom he had been living, decided him to quit this capital where he had dreamed of gaining so much fame. He set out on September 26, 1778, after having written to his father who was in Germany, these bitter and heart-rending words: "If only there were *one* person here who had ears to hear, and a heart to feel." He returned to Salzburg, his native country, where he lived in a state bordering on misery. Worn out, discouraged and enfeebled, he would surely have yielded to despair had not an unforeseen circumstance revived his spirits.

Charles Theodore, the Elector of Bavaria, invited him to come to Munich, and write the music for a grand opera to be brought out at the Italian Opera in that city. It was a real stroke of fortune. Mozart left his home without delay, set to work at once, and at the end of several months brought out *Idomeneo*, an opera in three acts. This work, which resembled neither the Italian, German nor French schools, in its melodies, was considered as veritably a new beginning in the art of music. It was accorded a most favorable reception.

After this success, the composer received divers offers. The one made by the noble archbishop of Salzburg was entertained in preference to all others ; it admitted of the musician retaining his residence in his native town, which he loved devotedly. He entered at once into the service of his Holiness. This prelate, although a prince as well as archbishop, was coarse and avaricious. He thought by having a musician of genius about his person he but added one more servant to his present retinue. Therefore he tried to force him to eat in company with the domestics of the household. Mozart, feeling his dignity as an artist wounded, endured this humiliation for some time, but realizing fully his own worth, he finally broke the yoke so imprudently assumed. Recalling the reception that had been tendered him in his very childhood, by the members of the imperial court of Austria, he took the road to Vienna, and in 1782 brought out his ballet, *L'Enlèvement du Serail*, at the court of the Emperor Joseph II. This work met with a brilliant success, and now prosperity, or rather a peaceful contentment, seemed to fall to the lot of the composer.

The *Marriage of Figaro*, an opera in four acts, followed this ballet, and if this opera, so much admired at the present day, was not as well known then as it is now, we should remember, that nothing of all that had been heard in musical works before his day, could give the least idea of this matchless composition.

Wealth of thought, grandeur and richness in the concerted pieces, charming and original melody, full and varied accompaniments, all combined to the perfection of a work which marks an epoch in the artist's life, as well as forms an era in the history of dramatic music.

Exhausted by overwork, Mozart made a journey to Prague. There he remained awhile, and continued his work, but very soon returned to Vienna, the scene of his triumph. In this city he composed his last opera, masterpiece of all masterpieces, his immortal work, *Don Juan.*

Mozart, whom fate seemed to pursue, lived constantly in the greatest anxiety and trouble, spite of all the triumph and renown that he had gained. He could have been spared pecuniary annoyance, it is true, for the King of Prussia made the most enticing and advantageous proposals, to induce him to come to his court; but the artist, being fondly attached to the Emperor of Austria, declined the honor, disregardful of the advantages that would accrue from the acceptance, and sent a positive refusal. The Emperor of Austria, informed of Mozart's devotion to the court and to himself personally, expressed his gratification, or rather perhaps, his gratitude, by making the artist the most flattering promises. But kind words and promises from an emperor are meaningless and deceitful; once made, they are soon forgotten. The lot of the composer continued to be as uncomfortable as heretofore.

Forced work, the anxieties and cares of life, produced a serious change in Mozart's health. A trouble with the chest was brought on, and symptoms foreshadowed the end, which if not near was none the less sure. Mozart, urged on by his genius, devoted himself to the severest work with unparalleled energy, disregardful of physical suffering.

He brought out two operas in quick succession: one in two acts, called *Cosi fan tutte*, and the following year another, that pleasing composition, full of grace and sparkling ideas, and written in an entirely different style from his other compositions—the *Magic Flute*. These two operas, like those preceding them, met with the greatest favor from the public, and added much to the celebrity of the artist.

The strength of the illustrious master was

exhausted by too constant application and continuous ill-health. He could no longer write without becoming extremely fatigued; he suffered also from depression and weariness, which often threw him into the deepest melancholy His wife, all devotion, moved by the deepest tenderness and love, made fruitless efforts to divert him from an occupation which was evidently wearing him out; she even went so far as to take away his music-paper and his rough sketches, hoping thus to hinder his working. Vain efforts! Mozart, feeling the approach of his last days, became feverishly eager to transfer to paper the ideas which were crowding his brain and the emotions which filled his heart. A melancholy which he was unable to throw off darkened his whole life. One day, when he was more sad than usual, and absorbed in gloomy thoughts, an unknown person came to his house and asked to speak with him. "Sir," said the stranger, when he entered Mozart's presence, "some one has lost a very dear friend, and desires to pay homage to his memory by having an annual service performed, for which he begs you to have the kindness to compose a *Requiem*." Owing to the mood in which this solemn visitor happened to find him, Mozart eagerly accepted the offer.

"How long before you can complete the work?" asked the stranger. "In one month." "What price shall you ask?" "One hundred ducats." "Here they are," said the visitor, and withdrew.

Mozart set about the work. It would soon have been completed had his attention not been diverted for a time and a new direction given to his thoughts.

The Emperor Leopold II. was to be crowned King of Bohemia, and on this occasion there were to be splendid feasts and ceremonies. Mozart was commissioned by the managers of the theatre of Prague, to write the music of an opera of Metastasio, *La Clemenza di Tito.* This order coming in the month of August, 1791, was accepted by the musician; he straightway set to work, and one month later, on September 15, he wrote the last notes of the music of this almost colossal opera.

This prodigious effort of his creative genius tended to increase the disease, and consequently the feebleness of the composer. Eager to fulfil his engagements, he quitted Prague, and returned tc Vienna to resume the composition of the *Requiem* which had been momentarily laid aside.

As soon as she looked on her husband, Mozart's wife realized at once her impending loss; and then, under the pretext of having mislaid it, she concealed the score which he had already begun. This unexpected disappointment had a visible effect on the composer; but as his strength was gradually declining, he soon seemed to grow resigned. This condition did not last long, however, for to his sorrow and confusion, the stranger of the hundred ducats suddenly reappeared before him. Mozart made every excuse for his inability to fulfil his promise. "I know," said the stranger, "how impossible it has been for you to keep your word; but how much more time would you now need to finish your work?" "One month," replied Mozart. "Well! here are another hundred ducats. Adieu, for a month."

The wretched musician had not calculated how little of life was left him. He grew worse day by day, and on December 15, 1791, just three months after his return from Prague, where he had witnessed the production of the opera written in honor of the emperor's coronation, he breathed his last, not having yet reached his thirty-sixth year.

The *Requiem* thus begun was finished by a pupil of his named Süssmayer, afterward musical director at Vienna. We are all familiar with this admirable composition, and though it be a masterpiece, we cannot but regret that Mozart was not allowed to finish it with his own hand, for, without doubt, he would have imparted a peculiar charm to the character of this great work.

Mozart, though dying in the prime of life, left a very large and admirable collection of musical compositions. He composed in all styles, and always with success ; his works include dramatic music, symphonies, concertos, duos, trios and quartets, commonly called *musique d'ensemble.*

If fortune never smiled on him, if he was

doomed to suffer a wretchedness which amounted almost to despair, because unlike many artists of the present day he possessed no business faculty, knew nothing of avarice, a passion which too often torments less deserving musicians, it was because there was a certain dignity and independence in his character which were very becoming in an artist. If during his life-time he was not granted the satisfaction of seeing his genius recognized in France, it was because that nation did not possess the musical education that it does to-day, and because at that time it was impossible to disturb the routine in which we must confess in many respects we still move. It is hardly credible that while Germany was somewhat boastful and justly proud of such a composer, greeting him with cheers whenever he appeared, Paris should show itself utterly indifferent to that admirable opera, the *Marriage of Figaro*. It was withdrawn after the fifth performance.

Gluck had opened the way to the grand and the beautiful; he taught the lofty language of passion, inspired a love for powerful dramatic situations, made effective by full choruses. Mozart showed his superiority by greater variety of ideas, more flexibility of style and greater richness in concerted music; in every respect he stood unrivalled.

Music, as we see, had made immense progress and multiplied its resources. Musical instruments being more numerous and much improved, artists were able to produce greater effects; composition became more varied in form and harmony; the one subject of incessant study furnished more richness, both as regards song and orchestration; artists were at last sufficiently cultivated to be worthy interpreters of the works of the most illustrious composers.

At this time, when dramatic music engaged the attention of the public, with whom it was the favorite entertainment, religious music was not far behind, for men endowed with lofty poetical sentiments cultivated it with success.

Cherubini, who appeared in 1788 with an opera in three acts, called *Démophon*, the words of which were written by Marmontel, and Lesueur, a purely French artist, religiously preserved all the traditions of sacred music.

In Germany, composers of undoubted and incontestable merit did their share toward advancing the progress of the art.

Weber charmed and delighted all by effects that were unexpected, poetical and bewitching. He aroused the deepest sentiments of the heart, stirred up, at his will, the tenderest as well as the most violent passions, and by curious combinations excited the most varied emotions. The operas of *Euryanthe* and *Oberon* appeared like brilliant meteors upon the musical horizon, and *Der Freyschütz*, (Robin des Bois) hissed, treated with contempt, and finally dismissed from the boards after but nine representations, was revived later, and gained a great triumph, being performed three hundred and twenty successive nights. Weber died at the age of thirty-three years.

Beethoven, a dreamy and melancholy poet, dwelling always nearer heaven than earth, communing more directly with the ideal than the real world, inspires us with vague emotions, full of charm and sweetness. Deeply sensitive to the harmonies in nature, he adapted them with unparalleled felicity, and reproduced them with expression and feeling. His symphonies, all of a noble character, afford us even to-day inexpressible delight, while they arouse our deepest emotions and tenderest sensibilities.

While Germany riveted the attention of *dilettanti* upon herself, Italy, as in former times, was destined to take the lead in a progressive movement.

A young man, a veritable artist, appeared in the musical world. He invested art with a power and vigor hitherto unknown.

Gifted with wonderful ease of expression and an astonishing fertility of ideas, Rossini seemed to *improvise* operas, which were produced on the stage in quick succession. At first they were not favorably received; some were greeted with accents of disapproval and even hisses. But the young *maestro*, insensible to the attacks directed toward him and to the cries uttered around him, did not allow himself to be vexed or irritated by what he chose to regard as trifles. Convinced as he was of the real worth of his productions, he only replied to his adversaries and numerous critics by producing new works ever full of happy and brilliant inspiration, and displaying a vast and dazzling wealth of resource.

Paying little regard to the old-established forms in music, or those consecrated by custom, he resolutely broke away from them all, and, provided such effects could be produced as were intended, he did not hesitate to disregard all time-worn rules. Indifferent to the charge of ignorance as regards the principles of harmony, often launched against him, deaf to the fault-findings of those wishing to detract from his fame, and who were all the more fiercely opposed to him because he never deigned to reply to anything that was said, Rossini continued to compose unceasingly.

Full of inspiration, and sure of final triumph, why trouble himself about the indifference and apparent disgust of the public, when his operas were produced? It is strange, and worthy of mention, that his very best operas were those that received the least kindly welcome, even from the Italian people, usually so full of sympathy with that peculiar style of music.

In 1816, while at Rome, Rossini produced for the first time the *Barber of Seville* which he had written expressly for the theatre of that city. He conducted the orchestra himself, and at the same time accompanied some of the arias on the piano. After the brilliant overture, so familiar to us all, the chorus came on the scene, and sang their strains of marvellous beauty and sweetness; then Count Almaviva, wrapped in his gray mantle, addressed Rosina from under her balcony in the tenderest accents of love.

The plot thickened, and was unravelled in turn to an accompaniment of most delicious, sprightly and gay music; but the public far from being pleased, seemed rather vexed and annoyed at hearing such unusual melodies. The ears of the *maestro* were greeted with malicious comments and sharp ridicule, but having the greatest faith in the ultimate success of his work, he remained impassive at his instrument apparently indifferent to the signs of disapproval, and showed neither anger nor chagrin; there was even seen on his lips the playful and mischievous smile with which we are all so familiar. Without anxiety as to the ultimate fate of his opera, he foresaw that eventually welcome and applause would greet the lively, graceful and droll music of his merry *Barber.*

Rossini marched with firm step along the path he had just cut for himself. During a long series of years Italy was the scene of his success, and meantime his brilliant compositions remained a sealed book to France. But this inconceivable indifference could not last forever; the fame of the musician was destined to reach even beyond the Alps, to triumph over the narrow spirit of routine in France, to remove the obstacles usually opposed to the success of works of genius, and to crush out all rivalry, jealousy and blind hatred.

When Rossini's music was first heard at a French theatre, all the critics denounced the new comer. They attacked and abused the man who had been so bold as to attempt to bring the public, those self-styled *connaisseurs*, out of the rut where they had so long unconsciously lived. All in vain had operas of a remarkably novel style been brought out, at a very short distance from Paris; the applause that greeted them in Italy, though it reached our ears had failed of effect; we would not hear or learn anything new.

We must read the account of the first performances of Rossini's operas at Paris, to form any idea of the injustice and ill-will with which they were received, and of the violent attacks made upon them; twenty, even thirty masterpieces had given evidence of the *maestro's* genius; but they seemed powerless to carry conviction to the hearts of the audience.

This obstinate dislike, unfair criticism, and violent abuse was visited alike on the sublime beauties of *Othello*, *Tancred*, *Moses*, *Semiramis*, *Count Ory* and *William Tell*, that chief among all masterworks! Such blind prejudice recalls to our minds, spite of ourselves, the struggle of Mozart to gain a hearing for any of his compositions, and the hindrances that he met with, when attempting to bring his operas before the public, as well as this sad passage in a letter to his father:

"If I were only in a place where people had ears, and hearts, and some little knowledge com bined with a small amount of taste, I would will ingly laugh at the intrigues forming against me but I am in the midst of a world composed entirely of animals and brutes, as far as music is concerned."

Rossini was too gifted with the divine spark of genius not to triumph, sooner or later. The recitatives, accompanied by rich and brilliant orchestration, the fulness and grandeur of the choruses, the vast wealth of melody, the powerful dramatic element, the interest which is never allowed to flag for an instant, the duos, trios, and all the concerted music; in short, the fertility of resource, the grace, richness and vivacity, which this artist knew how to introduce so cleverly into all his works, could not fail eventually to win the approbation of even the most exacting critics. They were compelled to admire, and *did* admire. Rossini's music, performed almost to the exclusion of works of other masters at the theatres, modified or changed the musical taste of the public. Even the most celebrated composers could not resist the influence of this great *maestro*. His peculiar methods of harmonic combination and of instrumentation, as well as the originality of his melody and song, were extensively imitated, and even copied; the musical world was full of his disciples, or those who without studying strove merely to imitate this great artist; some of his followers, however, more skilful or more independent than the others, preserved their own individuality. All composers of that day felt the powerful influence of this great artist; in Italy, Donizetti, Bellini, Mercadante, the two Ricci, and even Verdi (although they are both dramatic and original); and in France, Boieldieu, Halévy, Herold, Auber, in fact the whole French school, as well as the German musicians, Meyerbeer, Mendelssohn, and many others. All these renowned composers have given evidence of originality, it is true, but showed, also, that they had not been indifferent to the impression produced on their mind by the works of this great and wonderful genius.

We cannot hesitate to admit that Rossini, up to the present time, is the greatest and most finished composer that the nineteenth century has yet seen. There is, indeed, a peculiar kind of music called the "music of the future;" Richard Wagner standing first among composers of this style of music, the man whose works are so rich in harmony, the author of *Tannhäuser* and *Lohengrin;* there are also others, gifted like Hector Berlioz, and many young musicians, with original talent, but Rossini's rival or even *equal* is yet to appear.

Here we will close our little history of music. Let us give the young artists of the new school free scope. The works they have already given to the world are remarkable enough to be regarded as a fair promise for further advancement; let us offer our heartiest wishes for their future renown and glory, and, in the meantime, commend their efforts as well as applaud the success they have already attained.

PART II.

CHAPTER III.

Sound; its cause, origin, production, and mode of transmission and propagation, according to physicists.—Its nature and origin, according to the author.

THE sense of hearing, and that of sight, occupy the position of preeminence among all the other senses in the animal economy. Lending each other mutual support, they are almost always in direct communication.

The ear, and by this word, I mean the entire system of hearing, is the medium through which we experience the liveliest sensations and keenest emotions. A word, a sound, an expression, excite the most diverse and opposite passions: love and hatred, calmness and rage, courage and fear, joy and sadness, jealousy and despair.

All the sublime virtues, such as piety, trust, resignation, are encouraged or repressed, made inviting or unattractive, through the sense of hearing. According to the poet:

> "Nemo adeo ferus est qui mitescere possit;
> Si modo culturæ patientem accommodet aurem."

Through the sense of hearing we are warned of danger, we form sudden and unexpected resolutions, we learn the feeling of duty, become familiar with sentiments of honor, nobility and grandeur. By means of this sense we gain ideas of encouragement and despair, gayety and sadness, happiness and grief; in short, we receive salutary advice, which gives a fitting direction to our thoughts.

Sound has long been a subject of deep and continuous study; it has undergone many experiments.

Its production, its propagation, its mode of transmission, its cause and origin, have been investigated with the greatest care, and have given birth to several theories.

One of these, generally accepted at the present day, however ingenious it may be, does not seem to me to be entirely in accordance with established facts, for it does not explain all the phenomena which we witness every day. Therefore I shall endeavor to set forth another, which to my mind is more satisfactory; but first let us ask:

What is sound?

According to physicists, sound is an impression which the ear receives, a peculiar sensation excited in the organ of hearing by the vibration of bodies, when this vibration can be transmitted to the ear by means of an intervening medium. Others hold that it is a vibratory movement which takes place in ponderable matter. According to the former, it is a quality of other bodies, especially of the air, which produces it when under the influence of agents, causing vibration. These definitions do not appear to me quite exact. They confound the effect with the cause; for an impression, a sensation, and a motion are nothing else than the necessary, inevitable consequence of some primitive cause. If these impressions, these sensations, these motions, are a result; what is then the cause of this result, what is the principle of this sensation, of this impression received by the organ of hearing? The air, they say again, is the generator of sound ; but if it were so, would not the air, always in motion, incessantly produce it? Most undeniably. And yet, how can we conceive of silence?

Yes, air is undoubtedly a vehicle of sound ; it transmits it, propagates it to greater or less distances, but it does not generate it any more than the eye generates light. If then the definitions given above do not account for the varied and marvellous effects which we witness, they are inexact and worthless. What, then, is sound? What its origin, its cause, its nature? Under what conditions is it produced? Under what influences is it manifested? I will endeavor to answer these questions.

Sound needs a medium, a vehicle for reaching the ear. This truth is recognized by all physicists, as well as the world at large. Air is generally this medium ; but gases, liquids, solids, and elastic fluids have also, in different degrees, the property of transmission. If all these bodies have this property, sound must be something in itself.

"When a solid body is struck," according to

the physicists, "the integral molecules of this body receive a sudden shock; they are displaced, they oscillate more or less rapidly, and begin to move. These oscillations pervade the whole of this body, as well as those which may be in contact with it, either directly, or by the intervention of another body; so that when one part of the sonorous body makes a vibration forward, it disturbs and pushes before it the contiguous molecules. These molecules agitate in turn those behind, set them in motion and themselves become quiet; the latter in turn put the molecules nearest them in motion, and like those before mentioned fall into repose, and so on; so that sound is propagated through the entire solid body, and when it reaches any point whatever, all the parts which it has traversed in turn are at rest, unless the first part may have recommenced vibrating. In this case, the new vibrations are propagated in the same manner following each other as before.

"The successive movements occur in the entire body as rapidly as communication is possible; but then we suppose alternate condensed and expanded layers to exist in the body where the vibrations take place; this arrangement we call sonorous undulations. If the body to which the vibrations are communicated be boundless, in every sense, sound is transported all around the sonorous body by means of circular waves, in a sphere whose radius is more or less extensive" (Beudant, *Elements de physique*). This motion, which is perceptible to the hand and the eye, communicates a certain agitation to the layers of air surrounding the body which has been struck, and by virtue of the buoyancy of the elastic fluid, it is transmitted by degrees, propagated to a distance, and becomes perceptible to the ear.

Bacon and Galileo had already proclaimed similar views; and Newton attempted to demonstrate by calculation how the transmission of sound depends on the elasticity of the air and of the conducting body. He observed also that the effect of a sonorous body consists in the condensation of the molecules of air nearest to it, and also in the direction of the impulse given.

These molecules of air, driven forward by the

impulsion of the sonorous body, rebound, owing to their elasticity, and separate at the same time from the sonorous body the molecules of air directly in front, so that each molecule of air is displaced, moving first forward and then backward; that is, there takes place, around this body, an alternate condensation and expansion of the air.

As is well known, there are various opinions among those *savants* who have made a study of the propagation and transmission of sound. Some contend that the molecules of even the most solid bodies are disturbed, agitated, and displaced by successive movements of expansion and compression; others, on the contrary, hold that sound is produced only by the displacement and vibration of the layers of air that are in immediate contact with the sound-producing body, when struck. All speak, indeed, of the transmission and propagation of sound in space or in bodies, but not one speaks of sound, in itself. All students, however, who wish to investigate, not the ultimate cause of all things, but the tangible cause of phenomena which they witness every day, should notice this omission.

Yet I understand perfectly that the layer of air which is nearest the body may be agitated itself, and even partake of the same motion as the body when the latter receives a blow, or is moved by friction, or some other cause; I understand perfectly that this displaced and repelled air is alternately condensed and expanded. Again, I understand that there would take place in the atmosphere undulations similar to those produced by a stone falling on a liquid surface, such as the surface of a pond; but I confess, in all humility, my reason refuses to believe in this expansion, contraction, and displacement that occur in the molecules of solid bodies, such as iron, steel, stone, brass, wood, etc.; therefore, on this point, I differ entirely from those *savants* whose theory has just been given. If the agitation of the air, its vibrations, its undulations, were really the causes of sound, we would have a right to inquire how it is, that very sudden vibrations of bodies are often very perceptible, when under other circumstances they are hardly heard. How, when capable of being produced simultaneously in one and the same body, they do not mingle in such a way as to give rise to the most confused sounds. How, also, an almost imperceptible shock, a scratching with a nail, a pin, or even the point of a pen, all of which are incapable of moving and disturbing the molecules of solid bodies, can produce a perceptible sound at the end of a long bar of iron or of wood. If the successive or simultaneous displacements of the molecules of matter really took place as an effect of such slight shocks and movements, we should, with justice, be terrified at the frightful floods of air which must be the consequence of the agitation of great masses of the atmosphere, such as take place in tempests and hurricanes, or at the chaos that would be the result of two or more sounds. What would be the result, then, were an orchestra numbering three or four hundred musicians to execute a symphony, or were a cannon to be fired off?

No body, however solid we may suppose it to be, would be capable of resisting these displacements, these molecular vibrations, these undulations of the air!

If the views heretofore given are not correct, then sound is elsewhere than in the molecular vibrations of solid bodies of the atmosphere, or of elastic fluids; let us then seek to discover its origin, essence and nature.

If we but recall the nature of imponderable fluids, such as heat, light and electricity; if we admit, what is accepted as truth by the whole world, that these fluids, either latent or apparent, are developed through the changes, or simply the modifications of all bodies, whether solid, liquid, or gaseous, changes produced by blows or friction, by chemical composition or decomposition; again, if we recall the readiness of almost all bodies in nature to be impregnated by one or more of these imponderable fluids, it will be very interesting to see if, when sound is produced in resonant bodies there is not an escape of some fluid, like magnetism, heat, electricity, light, which themselves are different manifestations of one and the same fluid; if, in a word, we may not be allowed to admit the existence of a *sonorous* or *musical fluid*, the name matters little.

If the production of sound, or of the sonorous fluid, be developed under the same conditions and circumstances as other imponderable fluids, if it follow the same laws, if, like them, it give rise to similar phenomena, shall we not have some ground, some reason, to suspect its existence? As for me, I unhesitatingly admit it.

My belief is based on an hypothesis; but in that I am like all physicists who thus base their own opinion and reasoning with regard to the existence of imponderable fluids. Effects are seen,—we attempt to guess at the cause.

Let us seek, then, to establish by comparison, the similarity or identity of the sonorous or musical fluid with the other imponderable fluids. The sonorous fluid, immaterial, imperceptible, physically speaking inherent in all bodies in nature, like the other fluids to which it is analogous, is manifested under peculiar influences and conditions.

Inert, latent, passive, as long as its equilibrium

is undisturbed, it becomes perceptible to the senses as soon as this equilibrium is broken. Whether it be developed by motion, friction, heat, compression, or chemical action, it is a wonderful and powerful agent, a generating principle of feelings, emotions and sensations.

Like the other imponderable fluids, in order to be made manifest, it needs instruments appropriate to its nature. Like them it has affinities more or less decided for various substances. Certain bodies are good or bad conductors for it, as is the case with the other fluids, therefore the construction of sonorous instruments should depend entirely on their fitness to conduct the sonorous fluid.

Certain laws, certain fixed rules, govern the manufacture of instruments adapted to conduct the electric, magnetic and caloric fluids ; rules and laws also regulate the manufacture of instruments intended for the manifestation of the sonorous fluid. The instruments produce various sounds, each affecting the ear differently, according to the nature and arrangement of their constituent parts. At one time the sound will be soft, velvety and mellow; again, on the contrary, it will be harsh, rough and disagreeable; it will be noise, for noise is the result of discordant sounds heard simultaneously and by nature opposed to each other.

If all bodies were equally good conductors, if at the same time, according to the opinion of physicists, the vibrations of their molecules were the true source of sound, if the displacement alone of the air generated it, all bodies set in motion would be expected to produce it, which is not the case. It happens very often, in short, that large masses of air violently agitated, produce only dull sounds hardly perceptible to the ear, when, on the other hand, under certain circumstances, the slightest cause gives rise to sharp and penetrating sounds. Hold a piece of cloth, shake it, and you will hear no sound, or at least, very little. Make a tuning fork vibrate in the open air or on a soft body, and you will hardly perceive the sound produced; but rest it on a solid body, that is, a body capable of being charged with sonorous fluid, and you will perceive at once sounds relatively very intense. In these different cases, the tuning-fork has had the identical vibrations, the undulations of the air have been the same, and yet the sounds have sensibly varied in their intensity and perceptibility.

There is something more, then, than mere vibration of bodies or displacement of air in the production of sound.

I continue the demonstration by analogy, without referring to the similarity of the sonorous to other imponderable fluids.

As to these fluids, liquid or elastic bodies may be good or bad conductors. Place a certain quantity of water in a glass, pass your finger around the wet edges of the glass, and soft, harmonious sounds will be produced. Combine two gases in fixed proportions, and you will produce noises and explosions that will break the glass. The derangement of molecules, the disturbance of their equilibrium, are nothing surprising, for in the same manner as heat accumulated in a body expands, modifies and melts it, just as the electric fluid breaks and destroys a Leyden jar which has been overcharged, or as too bright a light wastes and discolors the objects with which it is kept too

.ong in contact, thus the sonorous fluid accumulated in a body, may induce alterations, modifications and disturbances in its molecular construction which might destroy it completely. In what manner does sound bring about these changes? In the same manner as light discolors certain objects, as electricity decomposes certain matter and as heat volatilizes certain substances. We witness the effects, we can only conjecture the causes.

The sonorous fluid, like heat or light, can be reflected on the surface of objects upon which it falls; it follows, even in this respect, the same laws as the other imponderable fluids, that is, the angle of reflection is equal to the angle of incidence.

Hence it results that any one situated in the *focus* of reflected sonorous rays will hear another person speak even at a great distance, while individuals who are between the two *foci* will not hear the least sound. In the lower rooms of the Louvre at Paris there are two antique bathing-tubs, several *mètres* apart, and so placed as to afford two very distinct *foci* of sound. A person standing at one of these *foci* can hear distinctly whatever is said, even in a low tone, by any one at the opposite *focus*, while those in the intervening space do not hear a word, or even a syllable of what is spoken.

In one of the rooms of the Conservatoire des Arts et Métiers, at Paris, there are similar *foci*. Two persons in opposite corners of the room can distinguish perfectly, words pronounced by each other in a low tone, although individuals very near the speaker hear nothing. Are not these effects analogous to those of concave mirrors for reflected heat? No doubt on this point. The more numerous these *foci*, the more remarkable the effect. Sometimes but one syllable is repeated by echo; again, several may be heard. I had occasion to take a sail on the Lake of Luco, whose waters fall from a height over rocks and form the cascade of Terni, one of the finest in Italy, possibly in all Europe, and there after declaiming an Alexandrine verse slowly and in a loud voice, I heard it echoed back distinctly and without alteration. Some echoes change the sound, and impart to it a plaintive or mocking tone, just as certain bodies alter light and decompose it. Shall we attribute all these strange and marvellous phenomena simply to the vibrations or undulations of the air? I deny it, unhesitatingly, for were it so, every one standing between the *foci*, and in the midst of these undulations and vibrations, as well as those standing between concave mirrors, ought to hear what is spoken, or feel heat, which is not the case.

But it will be said, the experiments made on metallic rods, and extended cords, have demonstrated that the vibrations occur according to settled rules and principles, and that along the length of these bars or cords, there were formed what are called *nodes*, that is to say points where the vibrations ceased, and were no longer perceptible. I regard these experiments as proof of the correctness of my views on sound, and its analogy to the other imponderable fluids,—with magnetism in particular. In short, do we not know that magnets do not possess throughout the same magnetic force? Are we not aware that if a magnetic bar be rolled in iron filings, the latter are seen to adhere in abundance to the ends of the bar, forming bristling bunches, while the adhesion of the filings rapidly decreases as we approach the centre of the bar, where it amounts to nothing. That part of the surface of the magnet, where the magnetic force is imperceptible, has been called the neutral line, but the two points nearest the extremities, where the maximum of attraction is manifested, are called poles.

Every natural or artificial magnet presents, then, two poles and one neutral line. This we learn from works on physics.

If, then, the effects of the magnetic force are such as have just been described, if there are nodes, otherwise called two poles and a neutral line, in every magnetic bar, are not the nodes of metallic rods or extended cords, when put in vibration, analogous to them?

I continue, and here put the question: Why do not all the extended cords of a musical instrument such as a piano, violin, violoncello or guitar repeat together the tone or cry of the voice which utters sounds above them?

Why, of all the cords of the same instrument, do those only which are in unison with this voice produce sound? Did not the noise, the sound, the sonorous wave, as they choose to call it, which escapes from the mouth strike, disturb and agitate all the strings? Yes, but the sonorous fluid did not find in all of these its *Leyden jar*, nor in any the capacity for being charged with the sonorous fluid. This, I think, is the explanation of the phenomenon that can never be explained by the theory of the molecular vibration of the bodies, or of the undulations of the air.

If, then, certain bodies are good or bad conductors of heat, light, and electricity, the same rule applies to the sonorous fluid. The walls of a concert-room draped with woollen hangings, the curtains and furniture of an ordinary apartment, are poor conductors of sound; they absorb and stifle it, just as woollen or silk stuffs of dark color absorb heat and light.

The temperature of the atmosphere, the hygrometrical condition of the air, different hours of the day, have no slight influence on the development, quality and intensity of the sonorous fluid.

A concert given on a calm lake, on a moonlight evening, will produce a much more marked effect on the listeners than would be the case should they hear the same music in a concert-hall. The loneliness of the place, the silence of the night, will have, it is true, some spell on the mind and imagination of the listeners, and it is this spell, this silence, which enables them the more readily to receive the musical fluid. Give an open-air concert in the midst of a noisy and excited crowd, execute symphonies in the presence of an agitated audience, and you will produce only a moderate effect. The sonorous fluid, like electricity, drawn from every direction by the mass of those present, will *charge* but a small number of them.

If by the union of piercing instruments, by means of a large number of clarions, drums and trumpets, a great noise be produced, you will impregnate the organs of hearing of the assembled multitude with the fluid, and surprising effects will be the result.

Electricity, as we all know, possesses peculiar properties; it is positive or negative; that is, either *vitreous* or *resinous*.

Light is composed of seven different colors; heat has its special properties; so of sounds, it has its peculiar shades.

The rays of light decomposed by the prism, combined, two by two, assume certain colors; the primitive scale, made up of seven sounds, combined according to various rules, produces different shades of tone. Two harmonious colors will produce a tint agreeable to the eye; two sounds, or, to use a word which well expresses my thought, a ray of sound combined with another analogous to it, will produce a harmony grateful to the ear.

Rays of light, like rays of sound, can be combined indefinitely, and produce an extraordinary variety of effects. For instance, the red ray of the solar spectrum combined with the blue ray produces the tint of violet, which is agreeable to the eye. The blue ray, when mingled with the yellow ray, produces the green one, which is also pleasing: and so on with the others. The red ray, on the contrary, mingled with the green ray, produces a color disagreeable to the eye, a *false* color, as it is commonly called. Sound has also combinations grateful or unpleasant to the ear, concords or discords. Thus the note *C* sounded in connection with the note *E* a third above it, will prove an agreeable consonance. *C*, sounded with the *F* a fourth above (the interval whose reverse gives the fifth—that is if *F* is the lower note and *C* the upper, the interval is a fifth), produces also a consonance; on the other hand, *C* and *D* (an interval called a *second*) will be a dissonance; *C* and *B* (an interval called a *seventh*) is also a dissonance, and consequently both are relatively disagreeable to the ear. In electricity, when a given quantity of the positive fluid is brought in contact with more of the same fluid, repulsion takes place, while if the positive fluid be brought in contact with the negative fluid, attraction is produced. That is to say, the fluids opposed to each other when brought together cause attraction; those allied to one another, produce repulsion.

The same effect takes place in the case of the

sonorous fluid ; the ray of sound *A*, given in connection with the ray *B*, for which it has no affinity, produces a dissonance. Some sounds, then, attract, as it were, others repel each other ; hence spring agreeable or unpleasant sensations. From these facts practical applications might be made of which artists should avail themselves. The painter's talent consists in the choice and arrangement of colors, as the musical composer's talent lies in the arrangement and combination of sounds ; in their *accord*, in short.

Nature furnishes colors and sounds ; the artist uses and combines them in a way to attain a single purpose—harmony. If every ray of light has a distinctive character, if electricity produces its own peculiar phenomena, is it surprising that the sonorous fluid should have its special properties ? Certainly not ! The analogy, I might almost say the similarity, existing between all imponderable fluids, becomes more evident as we compare them with each other.

The vividness and intensity of the colors in the solar spectrum, like the intensity and peculiar properties of sound, are singularly varied and modified by external circumstances.

At noon-day, in fine weather, when the sun's disc is veiled by neither cloud nor vapor, the solar spectrum presents, clearly and brightly, the seven colors of which it is composed. In the morning or evening of a foggy day, when the sun has a yellow hue, as it is commonly called, the indigo and a portion of the violet are lacking, and the blue rays have become invisible. When the sun is red, the green and part of the yellow rays disappear in turn. These phenomena, brought about by the power of reflection or refraction of the *media* traversed by the ray of light, are especially perceptible when the sun is on the horizon; only the least refrangible rays then reach us: the others are thrown back into the upper part of the atmosphere.

The rays of sound, a term which I employ for want of a better to express my meaning, are also reflected, refracted, altered and modified by the *media* traversed, and for this reason they make certain impressions on our feelings and thoughts, and fill us with infinitely varied emotions, according to the nervous sensibility of the individual who receives them.

Sound, being an imponderable fluid, exists everywhere, and is inherent in all bodies. If it is in a state of equilibrium, then, like other imponderable fluids, it is, as it were, in a latent condition, that is, in a state of *silence;* but if this equilibrium be disturbed, like them, its presence is made manifest. For its development it needs to be put under peculiar conditions, and requires instruments, as a means of manifestation. These are to sound, what the Leyden jar is to electricity, what iron is to the magnet. What Grimaldi, Huyghens, Descartes, Malus and Fresnel say with regard to light, may well be applied to sound. "A very subtle fluid which pervades the entire universe, and is generated in every direction, by reason of a disturbance at any one point."

Man, through his own talents, has hit upon laws which govern the construction of instruments, and nature, aiding him in his researches, has furnished most valuable hints. The voice of the wind, the roar of the waves, the rushing of the tempest, the cry of animals, the human voice and language have revealed certain conditions of order and arrangement which are necessary for this construction.

However, I hasten to remark, although the physicists and the manufacturers of these instruments have settled upon the volume, intensity and extent of sound, which may be expected from them, although too, they may have determined the laws, and recorded carefully in books as well as taught by practice all the details of science and experience, still all instruments constructed on the same model, or according to the same principles and dimensions, or of the same material, are far from having the same tone. Musicians understand well how to distinguish the difference of tone between two violins, two flutes, two bassoons, etc., and this tone, peculiar to each of them, does not lie in the arrangement and putting together of all the parts, since each is arranged like the other, but, in their capability for being charged with the sonorous fluid, just as every electrical machine does not possess the same facility for being charged with electricity.

If, in this chapter, I have insisted on presenting details which are technical and dry, perhaps, for general readers, but still very important, it is from a desire to make my theory clear, and to establish the analogy and similarity of the sonorous to the other imponderable fluids: that is, to show that this fluid is cause, and not effect.

CHAPTER IV.

Capability of animals for being charged with sonorous fluid. —Vocal organs.

THE capability for being charged with sonorous or musical fluid is not confined solely to inanimate bodies in nature, for animate beings also have this property in a very high degree. The latter have beside, a great number of methods for manifesting sounds. But with all animate beings, the chief of all methods is, without doubt, the vocal apparatus.

It is more or less simple or complicated, according to the species; but it is not always an indispensable means for producing articulate sounds, for certain insects, which possess no vocal apparatus, produce sometimes very intense sounds by the friction of certain parts of the body.

Man, as well as all animals provided with a

vocal apparatus, has no other means than this with which to produce sound.

In man this vocal apparatus is absolutely necessary for modulating sound and uttering words. With regard to him and other animals which possess it, it is quite complicated, and its constituent parts are very numerous. Without entering into all the anatomical and physiological details, that would hardly be intelligible to the greater portion of readers unacquainted with the science, I may merely mention a few of the leading principles, instead of totally ignoring the whole subject.

The emission of the voice, and of speech, cries, and the infinite variety of articulate or inarticulate sounds, are not confined, as some authors seem to think, to the opening of the glottis, and to the movement of the vocal cords.

The respiratory organs, wind-pipe, larynx, pharynx, cavity of the mouth, palate, teeth, lips, tongue, and nasal passages, fulfil important functions in connection with the voice, and each, in its own way, assists in imparting fulness, force, depth, accuracy or weakness to sound. In a work by

Geoffrey St. Hilaire, published a long time ago, an opinion was set forth, the truth of which is generally recognized even to-day. "It is an undeniable fact, and supported by abundant evidence as well as admitted by those even who maintain the larynx to be the seat of the voice (though they agree that it is but the principal organ used for emitting it), that every part of the respiratory organs is employed in giving utterance to sounds. I admit that the larynx and glottis have an important place in the phenomenon which we are now considering, but I positively deny that they are the only *essential* organs. The bones, muscles, nerves and cartilage contribute respectively to the formation of voice and speech."

Analyzing the office that belongs to each of these, the physiological savant considers the thyroid (Adam's apple) as a sonorous body. It fills the same place in the vocal system as that of a "sounding board," in pianos and other instruments. On this assumption, Geoffrey St. Hilaire attributes the difference of tone between individual voices, to the difference of size, form and tissue of this organ in each individual.

The thyroid, of very delicate consistency during the years of infancy, gives a sharp tone to the voice. Later, having become firmer in texture, though still maintaining its original condition of cartilage, it imparts more clearness, fulness and extent to the voice.

At the age of puberty, the organs become stronger, the muscles are developed, and the voice *breaks*, as it is called. Then the muscles being more firm and powerful, keep the thyroid in constant motion, but eventually it becomes almost solid, and is then able to resist their action. By this time, being covered with granulations, it becomes ossified, loses its elasticity, and renders the voice deeper and more harsh.

These different conditions and modifications explain why the voice, in the case of women, children and *castrati*, whose muscular system has never been fully developed, is high and clear; that of adults deep; and that of old men, thin, sharp, and *out of tune*, as it were. The voice becomes hoarse, when the sonorous body has lost its elasticity, or when it has not been carefully managed, as is the case with public criers and street hucksters.

Being careful of one's *instrument,* as singers express it, is, trying to make use of precautions to prevent the too rapid ossification of the thyroid; to abuse it, as public criers do, hastens this development, and incurs the danger of bringing it prematurely into a state similar to that peculiar only to old age.

In assigning such an important rcle to the thyroid, Geoffrey St. Hilaire seems to me to make it of too much consequence, for the glottis, vocal cords, wind pipe, cricoid and arytenoid cartilages which form part of the larynx, can with justice lay claim to a full share in the phenomena which go to make up the voice. The same is true of the interior of the cheeks, of the teeth, tongue, lips, and nasal passages, all of which contribute powerfully to the manifestation and modulation of the voice, and consequently of speech.

Song, too, is capable of endless variety of

shading dependent on the separation of the jaws, the movements of the muscles which stretch or relax the vocal chords and move the cricoid and arytenoid cartilages more or less; in short, cause numerous changes in the mutual relations of all these organs.

From the preceding remarks, the following practical suggestions may be arrived at: if we wish the voice to retain its freshness and liquid tone, and would impart to it flexibility and lightness, we must carefully manage the action of the muscles attached to the bony or cartilaginous parts of the vocal apparatus. The more the muscles are kept active and free, in other words, the more easily they can be contracted and relaxed, the greater will be the flexibility of the voice. The teacher who would give singing-lessons, should order that the muscles of the cricoid and thyroid, as well as arytenoid, and those connected with them, be exercised gently. He ought to devote special attention to having muscles which work in opposition to these, kept at rest, or at least exercised very moderately.

This regard to the management of the muscles, and these precautions, will go far toward improving the voice, and will add charm and power to the music that it gives forth.

The effects produced by sound, or rather by the sonorous or musical fluid, are far from being fleeting or transient. This fluid, acting directly upon the nerves, and in perfect harmony with the solids and liquids that pertain to the whole animal economy, often impregnates them to an incredible degree. Let us recall the effects produced on our whole being by listening to music that deeply touches our feelings, particularly the trembling of our limbs, the agitation of our nerves, the emotions of pleasure or sadness, of calmness or joy, caused thereby; also the sensations we experienced, the sentiments and passions aroused in us, and we will have a proof of the lasting effect of the action of the sonorous fluid on our organs. No—as Grétry says, "The musical sentiment is not fleeting, like a substance enclosed in a vase, and which evaporates when coming into contact with the air." It has, on the contrary, a deep and lasting effect. What lover of music has not often been tormented and harassed to some degree, by the recurrence to his memory of certain airs which, spite of all he may do, cannot be driven from his thoughts. Constantly mindful of them, they vex, absorb, torment and irritate him. They perpetually recur, and it is impossible for his attention to be completely diverted. He recalls the variety of shading, but often, when trying to reproduce them with his voice, they vanish from his memory, and he fails to be able to give them audible expression.

Besides, in listening to a succession of chords, do we not foresee, do we not have a presentiment, as it were, of what is to follow? Do we not experience a pleasurable sensation of *anticipation*, if I may be allowed such an expression.

Whence arise these varied phenomena? Are they the effect of the memory alone? It may be, as regards those we have already heard, but it cannot be the case with those which do not exist, and which we can only conjecture. In such instances there is a physiological, magnetic, and possibly electric effect, but the effect of the sonorous fluid can also be detected. Our organs are imbued with it, or they are in a condition to become so.

Other analogous effects occur every day, without our being able to explain their cause, otherwise than by the presence of a fluid inherent in our organs. Remain for a long while within hearing of a continued noise, and you will retain this noise in your ears for a longer or shorter time after it has ceased, according to your sensibility, and by this word I mean the aptness of these organs to receive or develop the fluid just mentioned.

Travel in a carriage or on a steamboat for some hours, or rather for some days, and long after your journey is over, you will still feel the motion and hear the noise to which you have been subjected. After having been long on the water, you will seem to perceive an oscillation not only in liquids but in solids, as if an *aura* surrounded all bodies. Is it not the fluid still within us that preserves and keeps alive these impressions?

In all that we call antipathy and sympathy, can we not recognize again, beside the existence of the magnetic, that of the sonorous fluid? We often see a man for the first time, whom we do not know, have never even heard spoken of, and with whom we have had and are perhaps never to have any connection, and yet we feel a lively interest in him; an indescribable *something* attracts or repels us; in a word, we experience feelings of sympathy or antipathy. And why should there be any attraction or repulsion between this man and ourselves? It is because the magnetic, like the electric fluid, acts positively or negatively; it either attracts or repels.

It is a singular fact, too, that if we make the acquaintance of this same individual, if we hear him speak, and exchange a few words with him, both the sound of his voice, and intonation of his speech, will perhaps influence us straightway to change our opinion as regards him. His musical fluid, finding its harmonious counterpart in us, will cause our antipathy to give place, if not to a sudden sympathy, at least to a feeling of kindliness. Do we not observe similar phenomena every day in our social life or at the theatre; are we not very often full of enthusiasm over some operatic singer for whom we had but little sympathy before listening to the tones of his voice. If we deny the existence of the sonorous fluid, is it not inexplicable why under certain circumstances we feel an urgent need for hearing music, or singing, or simply making a noise of some kind? Why under other circumstances do we prefer silence, and listen with the greatest indifference to the musical composition which we once so warmly applauded and which at any other time than the present we would applaud again? Why this change, this anomaly? It evidently depends upon the favorable condition of the organs for receiving or being charged with the musical fluid. This condition is the real reason why on certain days singers give us false tones, spite of all their efforts, zeal and good-will to sing correctly, and however good the condition of their vocal organs may apparently be. They are however the same organs to-day as yesterday; their mutual relations physically are the same, but their condition is not favorable for receiving the sonorous fluid. If the vibrations of the air and of musical instruments are the cause and origin of sound, they ought constantly to produce the same effects, emotions and sensations, under the same conditions of time and place, which is, in reality, not the case. The Leyden jar and the electrical machine always remain the same, but they do not always produce similar effects, as regards power and force, with an equal degree of sensibility.

From all these facts may we not reasonably conclude that by regarding sound as an imponderable fluid, and our organs as means of manifestation, we can perfectly account for these wonderful, strange, and hitherto unexplained phenomena which we witness daily; also, that there is a strong and close bond between us and those bodies surrounding us, which is felt and realized in a multitude of events connected with every-day life; that this bond, call it sympathy, antipathy, magnetism—the name matters little—is manifested every day, every hour, every second, bringing in its train wonderful and extraordinary effects; that the existence of the sonorous or musical fluid, like heat, magnetism, and electricity, is established by the analogy in their mutual relations and in the consequences that arise from them. New experiments will one day throw a more brilliant light on this scientific point. In the meantime we must admit that sound, a pure and simple emanation, a physical characteristic of all bodies, a special fluid, or simply a modification of other imponderable fluids pervading all nature, always capable of being manifested, can, like heat, light and electricity, be combined and modified in a thousand and one different ways, so as to bring about the most curious and interesting results.

The combination of sounds among themselves, with occasional intervals between, has received the name of music, and this art, or science as we should call it, has a great power over the whole animal economy, and an influence that has been noted and acknowledged even from earliest ages.

Let us cast a glance, then, at the phenomena that it produces; let us study its effects, in order to use them, if we can, to assuage human suffering.

This study, I know, may appear, if not ridiculous, at least of no use, to certain minds, which are exclusively engaged in seeking out pharmaceutical and chemical means for treating diseases. But I hope that the truly philosophical minds will not regard the subject as entirely undeserving of attention. With Plato they will say, "Music, that perfect model of elegance and precision, was not given to men by the immortal Gods with the sole view of delighting and pleasing their senses, but rather for appeasing the troubles of their souls and the sensations of discomfort which imperfect bodies must necessarily undergo."

CHAPTER V.

Effects of music on animate and inanimate objects.—Its influence on the physical organization.

AS we have already seen, all races, from the most remote ages, have devoted special attention to music. It was subject to perpetual changes, while keeping pace with the various transformations among the different nations. But, on the other hand, producing everywhere similar results and phenomena, whether on animate or inanimate bodies, it has always been an object of study and and wonder. Hence, a marvellous power was attributed to it, and it was the source of a multitude of fables in which exaggeration and imagination played the chief part. But all the stories and fables to which it has given rise, however improbable they may be, nevertheless bear witness to its importance and its power of fascination.

Among the Greeks, Apollo invented the lyre,

and by playing upon it soothed the vigilant Argus to sleep. Orpheus tamed wild beasts by the bewitching tones of his voice.

Amphion erected the walls of Thebes by means of his songs. Terpander restored a rebellious people to their allegiance through his melodies, and Tyrtæus aroused a whole army to action by the sound of his flute. The legislators of antiquity utilized music as a method of government, and the philosophers and sages of Greece were the first to recognize the fact that it had a salutary influence on the prosperity and moral welfare of the people.

Plato did not hesitate to say that no change can be made in music, without a similar one being made in the state, and he claimed that tones could be selected, capable of arousing malice, insolence, and their opposites.

Aristotle, Plato's antagonist, agreed with him on this point, and Polybius, in speaking of that musical race of Arcadia, contrasts the gentleness of their manners with the cruelty of the Cynetes, who neglected the cultivation of music. All have admitted that music had power to calm base passions, and bring noble ones into play. Everywhere, even among the most barbarous people, religious ceremonies and sacrifices were accompanied by songs or cries, according to the passions, customs and manners of the race.

The blood-thirsty Nero laid aside for a moment his fierce and cruel nature, to dispute for the musical prize in Greece. Surrounded by musicians, he seemed while in their midst to forget his implacable feelings of vengeance.

Amurat IV., that cruel fratricide, shed tears when listening to a performer on the psaltery, and pardoned those whom he had just condemned to death.

According to Galien, a certain Damon of Miletus played on the flute in the Phrygian mode, and roused some young men who had partaken too freely of wine almost to madness ; soon after, he modulated into the Doric mode, and restored these furious youths to the most perfect calmness.

Gibbon, in the last volume of his work on the Decline and Fall of the Roman Empire, observes that it is proved by experiment that the action of sound, while accelerating the circulation of the blood, affects the human frame more powerfully than eloquence itself. He then cites the following anecdote, contained in an account of a journey through England and Scotland:

According to the most ancient traditions, the bagpipe has always been the favorite instrument of the Scotch, since it was first introduced into the country at a very remote period, by the Norwegians. The larger one figures in their battles, funeral processions, weddings and on other great occasions; the smaller sized one is devoted to dancing music. Certain martial airs, called *pibrochs*, produce the same effect on the natives of the Highlands as the sound of trumpets does on their chargers, and sometimes even miracles are performed almost equal to those attributed to the music of Greece.

At the battle of Quebec, in the month of April, 1768, while the British troops were retreating in disorder, the commander complained to a staff officer of Fraser's regiment, of the bad behavior of his corps. "Sir," replied the latter with some warmth, "you made a great mistake in forbidding the bagpipes to be played; nothing animates the Highlanders to such a degree, at the hour of battle; even now they might be useful." "Let them be played as much as you please," answered the commander, "if that can recall the soldiers to their duty."

The musicians received the order to play the favorite martial air of the Highlanders; as soon as the latter heard the familiar tones, they paused in their flight and returned with alacrity to their post.

In the Peninsula war, Sir Eyre Coste, being firmly convinced of the attachment felt by the Highlanders for the music of their native country, gave them fifty pounds after the battle of Porto Nuovo with which to buy bagpipes, as a token of his satisfaction with their conduct on that day. (Anecdotes from Waverley Novels. Walter Scott's notes in the novel, *The Fair Maid of Perth.*)

Do we not also read of numerous and wonderful instances of the power of music in the history of our own country?

During our great and immortal revolution,

while the excitement was at its height, at that moment when Europe allied against us was pouring her countless armies over our frontiers, did we not see the whole nation march to combat and to victory cheered by the noble and patriotic tones of the *Marseillaise* and the *Chant du Départ*?

This magnetic power of harmony, as R. P. Kircher so emphatically expresses it, pervades all objects in nature. Inert as well as animate bodies, those that are rough-hewn and hardly defined or perfect and complete, all alike yield to its sway. The strand resounds beneath the breaking waves; fearful sounds issue from the tempest-beaten rock, and the trees of the forest, agitated by the wind, fairly roar.

Sound penetrates all bodies, and if music is performed in a room, theatre or church, it is no rare thing to hear the furniture, boxes, or altar, resound in unison with the instruments. The sound seeks its harmonious counterpart, unites with it, impregnates the molecules of the bodies where it accumulates, modifies them, and alters their mutual arrangement. A glass may be broken by the powerful voice of a singer, in the same way as a metallic rod may be melted by the current of the electric fluid, or expanded by the accumulation of a certain amount of heat. A pillar in a cathedral at Rheims trembles, it is said, at the sound of a drum or trumpet which is struck or sounded in the middle of the nave. Can these surprising effects be produced by the vibration of the air, I ask?

Is the undulatory movement powerful enough to put in vibration and displace the molecules of these compact, solid masses? Certainly not! If the glass be broken, if the church pillar be shaken, it is because the sonorous fluid acts upon them as electricity and heat act on the objects which they penetrate.

If from the inert, we pass to the animate objects in nature, we shall see an infinite array of curious and unexpected phenomena opening before us. We shall discover all kinds of sensations, emotions and passions. We shall witness most novel, varied and interesting sights. In one place we shall behold scenes of enthusiasm;

ecstasy or frenzy, in another, great agitation of mind, recklessness of spirit, and even convulsions of the limbs; in every instance, the effects will be extraordinary and wonderful.

Animals are not all equally sensitive. The tones and harmonies of music do not produce in each an equal amount of pleasure or pain; therefore the effects which result vary according to the species as well as according to the individual.

Noise serves generally as a warning to animals. It arouses the instinct of self-preservation, makes them aware of the danger that is threatening, and sometimes even gives them an idea of its extent. Some animals utter cries for the purpose of fascinating or frightening their enemies. The serpent charms by look the prey he wishes to devour; the torpedo benumbs and paralyzes by discharges of electricity the enemy which pursues him; and, in the case of other animals, their cries terrify and put to flight the foe that is hunting them down or lying in wait for them.

At the roaring of the lion, the horse and the camel tremble in the desert; the sheep and the dog shudder with terror when the wolf howls, and the instinct which causes this agitation, is, beside being the sense of self-preservation, a phenomenon essentially magnetic. If this be the case—and for my part, I am very much disposed to admit it—we can readily understand the effect of the grunting of the boar on the elephant, or of the crowing of the cock on the lion.

These extraordinary phenomena produced by *rough* sounds, if I may be allowed the expression, will become much more wonderful still, when they result from a series of harmonies, combined according to strict rules, that is to say—music.

Without speaking of the schools of music-mad dolphins, which as some poets relate, came to the surface of Lake Lucrino, to carry some flute-players on their backs to the opposite shore of the lake; without placing any faith in the stories of flocks being fattened by the notes alone of the shepherds who were tending them; without even recalling the tradition of those rats that, according to Bourdelot, danced, keeping time to the instruments, at the Fair of St. Germains, and, to use his own words, which he saw execute very complicated dances with the greatest precision ; without paying serious attention to these various fables, we will relate facts which may easily be verified should any one doubt their authenticity.

Be present when a bird is having a lesson on the bird organ, observe him carefully, and you will notice that at the very first sound he opens his eyes, becomes very attentive, approaches the bars of his cage, and the gentle shaking of his wings, as well as the trembling of his body, show that he is affected by the sounds to which he is listening. If his lesson be prolonged for some time, and his organs become impregnated with the musical fluid, you will soon hear him warble, in low tones, some of the notes he has just heard. By dint of many repetitions and numerous patient attempts, the pupil will attain such perfection, in a longer or shorter time, according to his musical capabilities (by this, I mean, his capacity for receiving the sonorous fluid), that he will eventually be able to repeat the air that you have played to him on the bird-organ.

The story of the spider which visited Pelisson in prison is familiar to us all; and Grétry in his Essays on Music laments the tragic death of a music-mad spider that one of his friends crushed. This little insect used to descend by its thread on to the composer's piano, as soon as the latter began to play. Bees that have escaped from their hive are easily brought back by the sound of cymbals, or of pans and kettles.

The horse, excited by martial music or the sound of the clarion, grows restive, runs, and plunges furiously into the midst of the fray, and the thickest of the fight.

Dogs are keenly sensitive to the sound of certain instruments. Military music makes some of them utter woful cries. With others the same effect is produced by the sound of stringed instruments. Grétry observes that these animals howl, particularly when the discords are long sustained, but never if the melody is simple, or if the rhythm is appropriate to the chase or to battle.

Richard Mead, in his dissertation *De tarantula*, relates the following circumstance: A violinist, while playing one day on his instrument, noticed a dog by his side; the animal was strangely affected by a certain tone; he howled fearfully and was apparently greatly distressed. The musician, struck by this circumstance, sustained the tone considerably longer than usual, and, at the same time, carefully observed the animal's attitude. The latter, became violently excited, uttered plaintive cries, and seemed in a state of agony. The musician, anxious to know how the experiment would end, continued playing the same composition, when the poor animal being thrown into a paroxysm of anguish, was seized with fearful convulsions, and, the music not ceasing, finally breathed his last.

In this very curious instance, the dog, I have no hesitation in saying, died as much from being struck by the sonorous fluid, as if he had been affected by a discharge from an electric battery.

This influence of sounds, or of the musical fluid on animals, becomes more plain to us, upon hearing of the experiment tried at the Jardin des Plantes, and recorded in a pamphlet which, luckily, has just fallen under my eyes.

"The 10th Prairial, of the year VI., a concert was given to the elephants in the Jardin des Plantes, and the effect produced was as follows: The artists were the brothers Rousseau, the elder Adrien, Guichard, Chol, Chlart, Devienne, Neitioffer, Félix, Delcambre, Frédéric, Lefebvre, Veillant, all distinguished musicians, and the majority of them connected with the Conservatory of Music.

"The orchestra, arranged so that it could not be seen by the elephants, was in a gallery above their cages, and the players were ranged around a trap-door, which was opened at the beginning of the performance. The two animals, male and female, were named Hans and Marguerite. They were allowed to occupy the two cages that constituted their abode, and could both walk freely from one to the other. Everything was ready, and a profound silence reigned; the trap-door was noiselessly opened, and the concert began by a trio in *B* major, consisting of airs with variations in moderate *tempo*, for two violins and a double-bass.

"Hardly were the first notes heard, when Hans and Marguerite, their attention being roused, ceased eating the dainties provided by their keeper. Soon they approached the place whence the sound issued. The trap-door opened above their heads; the queerly shaped instruments, of which the ends alone were visible; the musicians suspended, as it were, in mid-air; the invisible harmony, which they tried to reach with their trunks; the silence of the spectators, the impassiveness of their own keeper; everything, in fact, seemed to inspire them with curiosity, astonishment and anxiety. They marched around the trap-door while raising their trunks toward the opening, and from time to time, stood on their hind feet; they then went to their keeper as if seeking to be caressed, turned away again, apparently more and more restless, and looked at those about them, as if they feared being caught in some trap. But these nervous movements soon ceased, and then, yielding themselves up, without evident anxiety, to the delicious sensations occasioned by the music, they seemed conscious of nothing beside.

"The happy influence was especially noticeable at the end of the trio, which the performers closed with the dance music in *B* minor from Gluck's *Iphigenia in Tauris*, a music of a wild and strongly marked character, which imbued the animals, as it were, with its own peculiar rhythm. From their movements and their gait, now accelerated, now slackened, sometimes uneven and again regular, one would have said they were following the undulations of the song and the beats of the measure. They often bit the bars of the cages, grasped them with their trunk, and pressed the weight of their body against them, as if the space were too limited to enable them to give expression to their feelings, and they longed for more. Piercing cries and hisses at times escaped them. Are these caused by joy or anger? the keeper was asked. "Certainly not by anger," was his reply.

"The passion of which they gave evidence, soon calmed down, or rather gave place to another, when the air, *Oh ma tendre musette*, was performed in *C* minor on the bassoon alone and without accompaniment.

"The simple and tender melody of this romance, rendered still more touching by the melancholy tone of the bassoon, attracted them as if by magic. They walked a few steps, and paused to listen; then they placed themselves directly beneath the orchestra, flourished their trunks a little, and seemed absorbed with amorous emotions; it is noticeable that while this air lasted they uttered no sound, their movements were slow, measured, and partook of the gentleness of the song.

"But both were not equally fascinated. While Hans remained as indifferent and cautious as usual, Marguerite became quite impassioned and demonstrative. This dumb show assumed a phase of agitation and excitement, when the merry and lively tones of the air *Ca ira* was given in *D* by the whole orchestra. It might be inferred from their ecstasy and cries of joy, sometimes low, and again in a high key, but always with varied intonations and also from their hissing, and moving to and fro, that the rhythm of this air, of which the *tempo* is gradually quickened, was continually urging them on to more rapid motion, forcing them, as it were, to keep pace with itself.

"But happily the invisible power that had excited these creatures to such a degree could also calm their agitation, and the sweet harmony of two human voices repeating an *adagio* from the opera *Dardanus*, allayed their violent excitement, and they became quiet.

"These effects, however wonderful they may appear, ought not to astonish us, if we but reflect that the passions of animals, like those of human beings, have naturally an absolute rhythmical character, totally independent of all education and custom.

"Music may have the power of arousing and exciting the passions if we distinguish the movements which accompany such passions, and the accents which give evidence of their existence. It can change or calm the emotions, at will, by combinations expressing the time, order and succession of the various movements that contribute the outward evidence of these emotions.

"But nothing better proves this relation and close correspondence of rhythm and melody with the movements and accents called forth by passion, than the indifference of these elephants when, for the second time, the air *Ca ira*, was played, immediately after that from *Dardanus*; the key only being changed from *D* to *F*. It was indeed the same song, but the expression was no longer the same; the same harmony was still there, but it had lost its original force; the *tempo* of the piece was the same as before, but it was less lively, and the rhythmical accents were exchanged.

"Other airs were played, such as the overture to *Le Devin du Village*, which excited them to positive gayety; *La Charmante Gabrielle* plunged them into a species of languor; then *Ca ira* was repeated for the third time, but in *D*, as it was first played, when the same symptoms were manifested as at the first hearing, the same agitation, the same frenzy and ecstatic delight."

To bring to a close this account of the effect of music on animals, which might be prolonged to an indefinite length, I will relate a fact known to my personal experience. It occurred in the suburbs of Naples. While sitting under the shade of a great tree, I was humming the air of an Italian opera. Looking out toward the country I was admiring its great beauty, when I heard close by me a rustling of dry leaves that made me shudder. I looked toward the place whence the noise proceeded, and perceived all around me a large number of those small greenish-gray lizards so common in Italy. Upon my moving, these animals retreated. I thought no more about them, but began to whistle the air I was previously humming. To my great astonishment I saw my listeners, who had a moment before disappeared, reassemble around me. Then I watched them with special attention, and at the same time continued to whistle. I thought I detected in them something like pleasurable sensation, made evident by the movement of their sides, the trembling of their bodies, and the expression of their eyes. I studied them still more carefully, and endeavored to sustain the tones with still greater accuracy. The lizards being charmed, perhaps even fascinated, seemed to take such keen delight in these tones, that they apparently felt perfect confidence in me, and were no longer frightened at my movements, which were as gentle as possible, and allowed me even to bring my hand so near as to be able to touch them.

This first experiment was too interesting not to be again tried, and every time I had an opportunity to repeat it, the same result ensued. If my voice, which is not a fine one, but still is *true*, could have such an influence on these animals, usually so timid, how powerful would be a harmonious and sweet voice, such as we sometimes hear. The *British Review* for the month of August, 1865, contains the following anecdote: "In Hall's expedition to the Polar Sea, in search of Sir John Franklin, Sterry, one of the crew, endeavored to harpoon the porpoises that were sporting around the ships, but as these sly creatures were not eager to come within reach of the weapons, the clever sailor tried to attract them by whistling. The whales and porpoises, according to him, can never resist the fascination of this music."

If music, or rather the musical or sonorous fluid, produces phenomena such as we have just mentioned, upon animals, what will be its effect on the physical, moral and intellectual nature of man, a prey to such varied feelings, emotions and passions. Music will be a delightful pleasure and a source of infinite enjoyment to some; in others it will produce uneasiness, ennui, suffering and even distress; in all, without exception, it will arouse sensations and impulsive emotions of some kind.

"There are," according to Cabanis, "peculiar combinations of sounds, and even of single tones, that affect all the faculties of sense; these, by their immediate action upon the soul, arouse certain sentiments over which they seem to have special power, in accordance with the primitive laws of organization.

"Tenderness, melancholy, sombre grief, gayety, merriment, martial ardor and rage, can be sometimes aroused and again calmed, by songs of most wonderful simplicity; they will be so all the more surely, if the songs are very simple, and the phrases which compose them shorter and easily comprehended." No! there does not exist a being who cannot find its harmonious counterpart in nature, and cannot feel an affinity for certain sounds that meet his ear! What is more extraordinary, and proves still more clearly the existence of the musical or sonorous fluid, is that, according to Grimaldi, its influence is not perceptible exclusively to the organ of hearing. Its effects are sometimes seen in deaf people, or in those who stop up their ears; many of these feel a kind of fluttering at the pit of the stomach, or a very marked contraction in the throat.

Vanswieten relates the story of a young girl whose muscles became convulsed on hearing a clock strike, and the author of this work has, more than once, while seated at his piano, experienced spasmodic movements of the throat and muscles, when playing certain strains.

Grétry, that learned observer and expert on the subject before us, mentions a surprising effect of music on the heart and the circulation of the blood.

"I placed," said he, "three fingers of my right hand on the artery of my left arm, or on any other artery in my whole body, and sang to myself an air, the *tempo* of which was in accordance with the action of my pulse ; some little time afterward, I sang with great ardor, an air in a different *tempo*, when I distinctly felt my pulse quickening or slackening its action to accommodate itself by degrees to the *tempo* of the new air. After all this," he adds, "can it be said that the ancients were wrong in maintaining that music had the power of exciting or calming individuals of a healthy organization who were passionate lovers of this art!" I have tried this experiment of Grétry's on myself, and the results have often been similar to those obtained by the illustrious composer.

Berlioz, that musician so full of originality, so little appreciated, and so sharply criticised, relates in forcible and highly colored language, the effects produced on him by hearing music of which he was particularly fond. "Nothing in the world," said he, "could give an exact idea of the effect, to any one who has never experienced it. My whole being seems to vibrate; at first it is a delightful pleasure, in which reason does not appear to participate at all. The emotions, increasing in direct ratio with the force or grandeur of the composer's ideas, produce, little by little, a strange agitation in the circulation of the blood; my pulses beat violently; tears, which usually give evidence of the crisis of a paroxysm, indicate only a progressive stage, and greater excitement and agitation is to follow. When the crisis is really reached, there occur spasmodic contractions of the muscles, a trembling in all the limbs, a total numbness of the feet and hands, a partial paralysis of the nerves of vision and hearing: I no longer can see, and can hardly hear—vertigo—semi-consciousness—"

These effects on the organs are not the only ones that have been observed: varied experiences have already been related. The celebrated *cantatrice* Malibran, on hearing for the first time Beethoven's symphony in *C* minor, at the Conservatory, was thrown into such convulsions that she had to be carried from the room.

CHAPTER VI.

Effects of music on morals and the intellect.

THE phenomena produced by music on the physical constitution of man are certainly very extraordinary, but their effect on his moral nature and intellectual faculties are much more marvellous. Countless instances have been related by authors on music, and each day shows us facts which seem almost like miracles.

As is related in the Book of Kings, Saul disobeyed the Lord ; thereupon the Spirit of the Lord departed from him, and he was tormented by the spirit of evil. The king's officers asked permission to bring into his presence a young man who could play on the harp, that he might hear its tones when he was possessed by this evil spirit. The king consented, and David was brought before him ; each time that the tormenting spirit sent by the Lord took possession of Saul, David touched his harp, and the king was comforted and became calm again, for the spirit of evil left him (Ch. xvi., 14, 16, 23).

Timotheus, while playing on the lyre the air known to the Greeks by the names of *Ortias*, could either rouse Alexander to fury, and invite him to take arms at once, or could easily quiet him.

Pythagoras, seeing a young man transported with rage, on the point of destroying his faithless mistress, begged a musician to play some air in the Dorian mode, and the anger and excitement of this betrayed lover, giving place to the most perfect calmness, he renounced all plans for revenge.

Ulysses, according to some ancient authors, was indebted to the gentle and chaste songs of the musician Phenius, for the fidelity of his wife Penelope; and Clytemnestra only yielded to the importunities of Ægisthus, after the base murder of the musician Demadocus, the guardian placed by Agamemnon, king of kings, to watch over his wife's virtue.

Such extraordinary effects upon the morals

would need to be verified in our day, for, if the efficiency of the method employed in similar circumstances, were admitted as authentic, I know of more than one husband who would be eager to place one or even several flute-players as guard over his wife. But alas! "*Autre temps, autres mœurs!*" I very much doubt whether the virtue of wives and the honor of husbands would be protected in our time and age, by the harmonies of any music, however chaste it might be. The style that we listen to every day, seems to me, on the contrary, more likely to excite the amorous passions, than quiet them; and in Italy as in France, in England as in Germany, husbands know too well the influence of the voice of a young and beautiful singer, on their beloved wives, or the power of a skilful and sentimental instrumentalist.

Could it be otherwise? Love, this secret, deep-felt, and natural emotion, which leads all creatures to seek companionship with others of its kind, which brings them together for mutual enjoyment, almost always springs into existence under conditions of happiness and health.

Final end and aim of man, and of all animate creation, love subdues even the most rebellious natures.

Music, by preserving a perfect equilibrium and harmony in all the organs, and filling the heart with tender sentiments, makes those who are sensitive to it, susceptible of a happiness amounting almost to a sweet delirium. It is so natural to love, to enjoy, without effort or exertion, the sweet tones of an emotional composition this is so delightful and intoxicating that we yield to our inclinations before we are aware of it.

But music, unfortunately, does not always produce such sweet sensations, does not always arouse such tender passions, for sometimes, on the other hand, it produces effects that are followed by terrible catastrophes.

A young musician of Provence, swayed by the passionate emotions inspired in him from witnessing the representation of the opera *La Vestale* by Spontini, could not endure the idea of descending to our prosaic world on emerging from the heaven of poesy which had just been disclosed to him ;

he wrote to his friends of his intention, and after hearing this masterpiece, with which he was so completely fascinated, for the second time, thinking with reason that he had attained the maximum of happiness reserved for man on this earth, blew out his brains on leaving the opera house.

In every composition, from the beginning to the end, we naturally, that is, without reflection, anticipate the succession of the various tones and modulations belonging to the key in which it is written. There are several keys, quite varied among the ancients, as I have before said, which among the moderns are classed under two principal heads, the major and minor keys. The first is generally adapted to sentiments of gayety, pleasure and contentment; the second is suited to the expression of sorrow, pity, fear, etc.

All men are not equally sensitive to these different keys. Some prefer sad, melancholy or pathetic airs; others enjoy light, merry and graceful songs. This difference in the matter of taste, and this affinity with, or antipathy to, the sonorous fluid, is so well recognized, that every day, in concert halls or at the opera, persons are heard complaining of the uneasiness felt when they listen to chords of sad and melancholy music, for instance, while others, on the contrary, when they hear sweet and tender melodies, are filled with infinite pleasure and delight.

In the application of music to health and sickness, I cannot refrain from remarking on the value of the above suggestion; it should remind us that it would be dangerous for a hypochondriac to have to listen to very grave and serious music; instead of dispelling any feelings of depression, if not despair, it would rather give rise to them. The choice of the key and the general character of the composition is therefore of the greatest importance. For this reason the poet-improvisatore who recites or declaims, always chooses a key most appropriate to the subject he is treating of, when he is to have instrumental accompaniment, and a physician should understand and appreciate the constitution of the patient whom he is to have the care of.

Grétry, in his Essays in Music, insists strongly

on the importance of discrimination as regards the key and general style in musical compositions, and lays down a series of rules, on this subject, which seem to me useful for composers to know.

"The key of *C* major," said he, "is noble and frank; that of *C* minor is pathetic. The key of *D* major is brilliant, that of *D* minor is melancholy. The key of *E* flat is grand and also pathetic; it is a semitone higher than that of *D* major, and still does not in the least resemble it. By ascending again a semitone, we reach the key of *E* major, which is as sparkling as the preceding one was grand and melancholy.

"The key of *E* minor is rather sad, although it is the first minor scale in nature; that of *F* major is mixed; that of *F* minor is the most pathetic of all; the key of *F* sharp major is hard and sharp, because it is overloaded with accidentals; the same key in minor still preserves a little of the same hardness; the key of *G* major is warlike and not as grand as *C* major; the key of *G* minor is the most pathetic, except that of *F* minor. I pass on to the key *A* of major, which is a very brilliant one ; that of *A* minor is the simplest, least brilliant of all. The key of *B* flat is grand but less so than *C* major, and more pathetic than that of *F* major ; that of *B* major is brilliant and gay, while that of *B* minor is adapted to express sincerity and artlessness.

"In general all the minor keys are tinged with melancholy; they are most used for expressing abstract metaphysical sentiments, that is, in short, all that are not of a defined character, such as grief, melancholy, dissimulation, irony, etc. . . . . A key should be chosen analogous in character to that of the person; for, if you make an old man sing merrily in the key of *E* major, and a moment later a little girl is made to sing in the key of *C* major, I should say that the key of *C* was adapted to the old man, and that of *E* to the little girl. If you make a warrior or a joyful lover sing in the key of *E* flat, I should think that the recital of his exploits would close with a catastrophe."

These observations seem to me very correct, and the musician who follows these precepts most carefully, who wishes to adhere closely to nature, will consequently produce a more finished work, and one that will be better appreciated, not only by artists and *connaisseurs*, but also by those whose musical capacity is not very fully developed. The music of the great masters supports this truth. Beethoven would never have been able to write his Pastoral Symphony in the same key as that of the Heroic; and the latter could never have been composed wholly in the key of *E* flat.

These precepts, so eminently true, unfortunately do not serve as guide to all composers, and music is not always written in the key which the words or situations would seem to demand. Among the Italians especially, we meet with the most repulsive anomalies. How often in fact, do we not hear the most pathetic words accompanied by the liveliest kind of music, and the most solemn arias, which should be free from all flourishes, are overburdened with grace-notes and trills!

What a misconception to make a warrior or an old man breathe his last, while singing *runs*, and how dreadfully out of place it is to have a *prima donna* strain her throat in executing endless *fiori-ture*, just at the moment when she should only breathe forth stifled sighs, or short words interrupted by sobs !

If each key have its special character, the same may also be said of every instrument.

The bassoon is mournful; consequently it should be employed in expressing sorrow and pathos. The clarinet is suitable for the expression of grief; and if it is used for rendering merry music, the same is sure to be tinged with sadness. "If a dance were to be given in prison," said Grétry, "I should like it to be to the sound of a clarinet." The oboe suggests reverie by its rural tones. The flute is sweet and tender; it is best adapted to express the sweet delight of a happy and tranquil love. The trombone is deep and harrowing. The trumpet excites frenzy and martial ardor. The violin seems suited to express all the sentiments common to humanity, but the viola ought to be reserved for songs of a tender melancholy.

It would be a really curious study to define the

peculiar characteristics of each instrument; and this task, possible only to professional musicians, would not be barren of results useful both to composers and physicians. The former would learn how to assign to various characters songs suited to their special natures; the latter would find useful and unforeseen resources to enable them to restore their patients to health, or relieve those unfortunates who are a prey to suffering.

CHAPTER VII.

Effects and influence of music on the sick.

IF music has an undeniable influence on our organs and ideas, if it can arouse and calm the most violent passions, as well as the gentlest emotions, what a power it must exercise on man, when either in a state of health or disease.

The human body is, as it were, a noble instrument, which, by reason of its complicated mechanism, as well as the incompetency of the one supposed to control it, and of the musicians who play upon it, is exposed to countless changes and derangements. When perfect harmony exists, health predominates; when some of its parts are in an abnormal condition, there is a sense of discomfort; when the equilibrium is broken, there is sickness. Sickness, in fact, as I have already said in a previous work, *De la Santé des Femmes*, vol. 1, is only the resultant of opposite forces;

and the great secret of preserving health rests entirely in maintaining the complete harmony of all the parts composing the human instrument, and in the wisdom and foresight of the one who plays on it. Music, warmly recommended by the ancients as a therapeutic agent, has remained, nevertheless, in this regard, buried in culpable oblivion. However, the resources peculiar to it are very manifest, and we have reason to be astonished at the indifference of physicians and hygienists as to its use.

The instances of diseases treated, cured, or at least relieved, by skilfully combined sounds, are numerous and well authenticated and it would only be necessary to apply this method, to realize what great advantages could be derived from it.

Athenæus, Theophractus, Aulugellus, employed music as a cure for sciatica and gout. Cœlius Aurelianus mentions a flutist who, by playing in the Phrygian mode, *charmed*, as it were, the diseased part, causing it to palpitate and tremble, which, as already said above, was only the result of the accumulated sonorous fluid in the affected part.

It cannot be otherwise, for if the relief had been due to the trembling alone, dancing, singing, see-sawing, or any other movement, would have brought about the same result, that is, would have served to alleviate the pain—which was not the case, and which never could happen.

Rhythm and measure have their influence also. By giving a certain character and variety to melody, the animal economy is modified in a peculiar manner, whence arise marked effects and results. Nurses and mothers of families often dry their children's tears by singing merry or monotonous refrains, in even measure, and with regular cadence. I have often seen children weeping, without being able to discover the cause of their tears, and remarked that they suddenly became quiet on hearing a piano, a drum, or simply the noise produced by drumming on the window panes with the fingers.

They are often calmed by being rocked, or patted at regular intervals. In such instances, is the infant put to sleep by congestion of the brain alone? Possibly, when it is rocked for a long time ; but do not sound and rhythm play the chief part ? Magnetism and electricity are often efficacious in treating pain, or any physical ills ; why then should not the sonorous fluid, which is so analogous to these, have a similar power.

Baglivi expressly recommends exercise and motion for persons afflicted with gout. If they cannot walk, let them exercise their voice, by reading aloud, conversing with their friends, or singing.

Bonnet says he has known several persons suffering from gout who employed music as a means of relief for acute pain, with entire success.

Sauvages mentions a young man, one of his patients, who had been attacked with intermittent fever, accompanied by violent headache ; he could only be soothed by the sound of a drum. The patient's friends were obliged to beat the drum constantly in his room, and this noise, which was deafening to everybody else, brought wonderful relief to him, although he did not like music when in good health.

Very many diseases of the lungs and stomach

are wonderfully alleviated, by speaking, convers-ing, or listening to music.

Roger recommends this last mode of treatment for pulmonary phthisis; it is always, in his opinion, followed by perceptible improvement.

A physician of the last century used to ask his patients who complained of difficult digestion, whether they took their meals alone or in company with others, and also if they talked or were silent while eating; he then advised them to carry on conversation, in order to lengthen the time of the meal, and thus force them to masticate their food more thoroughly; at the same time he intended them to exercise their lungs and develop the sonorous fluid. In order to relieve the pulmonary, as well as the vocal organs, of the blood that fills them in consequence of constant and repeated acts of swallowing, a sort of gymnastic movement is needed, without which the functions of every organ would fall into disuse; but this movement is not entirely sufficient in itself. Speech and conversation, by imparting activity to these organs, force the blood to circulate, preserve equilibrium in the different parts, and thereby prevent the feeling of oppression, the flushes of heat, and heightened color in the face, all such frequent effects of rich dinners. Eat a quantity, drink a certain amount of liquid at a single swallow, talk very little, and your breathing will be short and labored, your digestion will be rendered more difficult. On the other hand, eat and drink the same amount, and carry on at the same time a lively conversation, even a discussion with those at the same table, and you will digest your dinner with ease.

If you listen to good music after the meal, although you may be less sensitive to its influence than at other times, you will feel all the better for it. Voltaire, that witty critic, when he said that our purpose in going to the opera was to promote digestion, proclaimed an established truth, the full meaning of which he undoubtedly had no idea of. Listening to good music is undoubtedly the best of the various modes of exercise so often suggested to literary persons leading necessarily a sedentary life.

Milton, a very skilful musician, spent a certain

time every day after dinner in singing or playing on some kind of instrument. Boerhaave, the noted physician, indulged in the same pastime after each meal; and the author of this work never digests better than when he listens after dinner to some opera or symphony.

These effects are not less perceptible on man in a state of sickness, than in a state of health. Ancient as well as modern authors give countless instances of cures effected by means of music.

Democritus informs us that the sound of the flute is a remedy against the plague. Thales of Crete delivered the Lacedemonians by similar means when attacked by the same scourge. Celsus points out different methods of influencing the minds of the insane, depending on the nature of their mania. To quote his own words. "We must quiet their demoniacal laughter by reprimands and threats, and soothe their sadness by harmony, the sound of cymbals or other noisy instruments." According to Cœlius Aurelianus: "In the treatment of madness, some physicians employ exciting music without any discretion, which may produce good effects when applied in the right way, but on the other hand, cause much harm in a great number of cases. It is said that the Phrygian mode, full of sweetness as well as vivacity, is very admirably adapted to those who are one moment overwhelmed with grief and the next thrown into paroxysms of rage; the martial Dorian mode suits those who are given to talking and behaving in a silly manner, and indulging in bursts of meaningless laughter, but it often happens that such tones throw the patients into a delirium of excitement that is a kind of trance, which gives rise to the saying that they are possessed with the divine spirit.

Galien recommends the employment of music as an antidote to the bite of vipers and scorpions, and Desault claims the greatest success in the use of it for the treatment of hydrophobia.

The records of L'Academie des Sciences, at Paris, contain instances of cures that are really extraordinary; we may mention a few of them:

"An illustrious musician and noted composer was attacked by a fever, which constantly increased in violence, so that the paroxysms followed each other in quick succession. The seventh day after his first attack, he lapsed into a very violent and continuous delirium, accompanied by cries, tears, terror and perfect sleeplessness. The third day of his delirium he asked if he might hear a little concert in his room; his physician consented, but with many misgivings.

Bernier's *cantata* were sung. As soon as he heard the first notes, his countenance became calm, his eyes assumed a quiet expression, and the convulsions ceased entirely; he shed tears of pleasure, and the fever left him while the concert lasted; but as soon as it was over, he relapsed into his former condition. His attendants could not do otherwise than continue the use of a remedy, the success of which had been so unexpected and so complete. The fever and delirium always subsided during the concert; and the music became such a necessity to the patient that he had the person who watched with him at night sing continuously. Ten days of this *music treatment* effected a complete cure, without the use of any additional remedies, except that the patient was bled once in the foot; this had been tried before in the early part of his illness.

A dancing-master of Alais in Languedoc becoming excessively fatigued during the Carnival of 1708, owing to the demands of his profession, was attacked by a violent fever. On the fourth or fifth day he fell into a lethargy, and for a long time all attempts to rouse him proved fruitless. At last, instead of recovering consciousness, he passed into a condition of dumb but raging delirium, during which he made constant attempts to leap from his bed, trying to strike those restraining him with his head; though maintaining perfect silence, he obstinately refused to take any of the remedies offered him.

The mayor of the town, on seeing him in this condition, thought that possibly music could restore his deranged mind, and therefore proposed it to his physician. The latter did not disapprove the plan, but he thought to put it in execution might excite ridicule, which would be infinitely increased, were the patient to die during the application of the remedy. A friend of the dancing-master, knowing how to play on the violin, took the one owned by the patient, and played some airs familiar to the latter. He was thought more insane than the patient himself, and was treated insultingly, when all at once the sick man sat up in his bed, as if agreeably surprised, and endeavored to beat time with his arms, but as he was held back by force, he could only signify, by nodding his head, the pleasure he was experiencing. By degrees, however, the attendants relaxed the force with which they were restraining him, and allowed the patient free movement; he soon recovered his reason. At the end of a quarter of an hour he fell into a deep slumber, and while thus sleeping, the crisis of the disease passed.

A fact which I witnessed in the course of my medical studies, deserves also to be mentioned. In 1832 one of my relatives very dear to me, a doctor of medicine and devoted lover of music, had an apoplectic attack. Being hastily summoned, I was the first to prescribe for him. In spite of vigorous treatment, the patient did not recover consciousness until two days after the shock. One half of his body was paralyzed; and his speech much affected; some days passed, full of anxiety, and a fatal result was feared. Meantime his condition became improved, and he could to some extent express his desires and make known his wishes. The amelioration became more perceptible every day, but he still suffered from drowsiness, restlessness, and occasional paroxysms of fever. The patient, while in this state, expressed a desire to hear some music; his daughter was asked to sit down at the piano, and she consented, though not without great anxiety. Her father, who had up to this time been drowsy and silent, seemed to rouse up at the very first notes. An expression of delight overspread his countenance. His lips moved, he smiled, and joining his hands together, he seemed to drink in the sounds proceeding from the adjoining room with unspeakable delight. As long as the music lasted he was perfectly calm, apparently in a state of comfort, which continued till evening, when the improvement was less marked, although there was no sign of his condition growing worse. This first experiment had brought too much pleasure and relief to the patient, not to have the family eager to repeat it. On the next day I was present at the second trial, and experienced surprise and astonishment at the results I witnessed. Music was given daily, for his benefit. The grave and serious style of that played the first day, gave place to compositions of an entirely different character, and the piano was brought into the sick-chamber. The improvement was more marked every day, the convalescence continued without interruption, and the cure was rapid and complete.

These very interesting facts are worthy the attention of observers. They bear testimony in favor of music considered as a therapeutic agent; they also give the extent of the advantages that may be derived from it.

The action of sound, and the influence of the sonorous musical fluid, seem chiefly useful in nervous diseases, especially among women.

Hypochondria, hysteria, and even epilepsy have

been alleviated in an incredible manner by the harmonies of a music adapted to the patients for whose benefit it is employed.

Quarin relates an instance of epilepsy cured by music, where chance alone pointed out the remedy. These are the details: "One day the patient having been listening to music when she felt the epileptic fit coming on, suffered only the symptoms. Every time afterward that she felt the approach of the paroxysm, the young girl was placed so that she could hear music; and nature," says Quarin, "being thwarted, as it were, in its perverted tendencies, lost, at last, the habit of convulsive movements."

An analogous fact is related by Roger. A young lady belonging to the department of La Drôme suffered from a nervous disease much resembling catalepsy. The sound of the violin relieved her in a surprising manner, and if she had the good fortune to hear it before the paroxysm was really upon her, she was saved from it entirely.

Pomme mentions several cures brought about

by this means, and Pinel, speaking of a young girl, says: "During the attacks, the sense of hearing far from being deadened, seemed to have acquired more keenness. A skilful musician played on the violin at the patient's side, during her paroxysms; although she then appeared insensible to the charms of the music, she was so strongly affected by it, that she admitted, after having recovered entire consciousness, that the music had thrown her into a state of rapturous delight."

Désessarts relates a fact witnessed by himself: "In 1801 Dr. Duval, member of the Medical Society of Paris, was summoned to visit a woman sixty years old. Since the age of thirty she had been an invalid, having never fully recovered from a sudden fright experienced at that time. She was subject to veritable catalepsy; convulsions took place every year, at the same time, and it was during one of these paroxysms that Dr. Duval first saw her. He tried several remedies in vain. At last it occurred to him that hearing the notes of a clarion might prove a useful stimulant, fitted to the patient's natural gayety. She was not in the least affected by it. The Doctor had her then listen to familiar songs, and the way in which she moved her lips led him to think that her interest was aroused. The physician, on being informed that she was when in health passionately fond of hearing certain kinds of music, one day brought with him a musical friend who played several airs on the clarinet. These airs being unfamiliar to the patient she was not affected by them; but, on the other hand, was much moved by hearing Christmas carols, and other airs, which seemed particularly to suit her fancy. The impression once made, became more vivid day by day; she first began to mark time with her hand and head; on the fourth day her body recovered its flexibility; she was able to exchange her bed for a chair, thereupon was commenced the song *Confiteor*, when she rose and joined her hands as if in prayer.

"Dr. Duval now took her by the hands, made her dance, at first gently, and then in a more lively manner. The limbs, that had remained in a cataleptic state for four days, obeyed her will: the patient was able to accompany the Doctor to the end of a long hall, and on the following day, as far as the street, descending three flights of stairs. From this time she could attend to her customary occupations."

A fashionable lady, about twenty-seven years old, endued with extreme nervous sensitiveness, passionately fond of music, an art which she understood and keenly appreciated, told me, some days since, that when listening to the tones of an orchestra, or even of a piano, she became, at first, very excited, felt her limbs tremble, her heart beat rapidly, and her very flesh creep; then she experienced throughout her whole body a vague sensation of ease and comfort, as if some *fluid* (to use her own expression) were penetrating every organ. A perfect calm and sense of contentment amounting almost to delight, had succeeded her first feeling of excitement, her trembling, and the creeping of the flesh. Her thoughts were all happy ones; her heart, filled with the warmest sentiments, longed to give expression to its tender emotions.

This same lady said to me again, although she did not know of the work that I am engaged on, at this moment: "If I should have the misfortune to become insane, the best remedy, in fact, the only one from which any happy effect might be anticipated, would certainly be music."

Mental diseases are, under many circumstances, relieved or cured by a style of music suited to the peculiar kind of insanity or mania.

"A physician, one of my intimate friends," writes Bourdelot, in his History of Music, "having been summoned to visit a young woman, made insane by the faithlessness of her husband, succeeded in restoring her to reason, by bringing her into the presence of a number of musicians. The latter, placed behind a curtain, played, three times a day, a kind of music suited to her condition. This remedy, continued for eighteen days, produced the happiest results."

One of my friends, a young artist, lost his reason, six months ago. He is never sane except while listening to singing, or humming to himself airs from Italian operas.

Should we not commend, every day, the happy idea of employing music in our insane asylums, for the cure, or at least the relief, of certain kinds of insanity.

It is especially in mental diseases, for sadness depression, despondency and *ennui*, and as an antidote to disgust for life, that music may be employed most successfully. It performs wonders, in cases of a vague longing for new excitement; gratifies those aspiring for mental enjoyment, and cheers all who suffer from loneliness of spirit.

A divine emanation, perhaps, source of inexpressible delight, it arouses courage, affords new food for thought, recalls hope, and inspires most agreeable fancies.

I might give countless instances of mental diseases cured, or at least alleviated, by music, but prefer to confine myself to copying here a page full of eloquence addressed by George Sand, one of our most distinguished novelists, to Meyerbeer, one of our renowned musicians, *à propos* of the opera, *Robert le Diable.* I also appeal to the memory of my readers, each of whom will probably be able to relate a story or chronicle some successful experiment.

"Music has much greater influence than decorations. A complete melody is not needed; simple modulations are all that are required to make dark clouds pass over the face of Helios, or to touch the azure vault of heaven, to stir up the volcano, to make the Cyclops roar in the bowels of the earth, to recall the moist breeze, and make it fan the trees withered by terror.

"Alice appears, the weather is serene, nature gives us wild and primitive harmonies. All at once the witches dance frantically in circles before her, the earth trembles, the grass withers, the subterranean fires come forth from all the pores of the groaning earth, the air grows dark, and weird lights illumine the rocks. But the witches' vigil penetrates the almost inaccessible recesses of the caverns, nature is revived, the sky grows clear, the air becomes fresh, the brook resumes its course which has been interrupted by fright, Alice kneels down and prays.

"In this connection, and spite of the length of

this digression, I must relate to you, my master, a fact of personal experience, and for which I have always intended to express my gratitude to you.

"Two years ago I went into the country, in the middle of winter, and passed there the two saddest months of my life. I suffered from dreadful depression, and during the attacks, almost lost my reason. I seemed possessed by furies, demons, serpents, and everything that was incident to the vigil in your opera. When the crises, as happens in the regular progress of all diseases, began to pass off, I had an infallible means for hastening their course, and thus in a few moments arriving at a blissful state of repose. I had my nephew take a seat at the piano: he was a handsome curly-headed, rosy cheeked young man, of a serious turn of mind, filled with a gentle dignity such as would befit a monk, having besides an almost impassible countenance, and blessed with even health. At a signal which he understood, he played my favorite music, that sung by Alice at the foot of the cross, such a perfect and charming picture of the condition of my feelings, the passing off of my distress and my returning hope. What poetic and religious consolations have fallen, like blessed dew, from these sweet and penetrating notes.

"The chaffinch on my white lilac forgot the cold of winter, and dreaming only of Spring and Love, began to sing as if it were the month of May. The flower, half-opened on my mantel-piece and unfolding its silk-like petals, shed a chaste perfume above my head at the sound of the closing chord. Then the aloe's leaves were lighted in my Turkish pipe, the hearth was illumined by an intense white light, and my nephew, patient as a machine, devoted as a son, recommenced the same phrase twenty times in succession, until he saw his dear uncle throw off the swan's-down wrap in which he was enveloped, and attempt the most graceful steps in the middle of his room, while throwing his cap up to the ceiling and sneezing for twenty minutes.

"How can I ever bless you, my dear master, who healed me so much more skilfully than a physician would. for you did it without causing suffering and without demanding any pecunlary compensation; and why should I believe, now, that music is an art purely for pleasure or mere speculative enjoyment, when I remember the surprising influence it had upon me, and that its eloquence was more convincing than all the philosophy taught in books?"

What a magnificent example of the power of music in relieving mental diseases! What an encouragement for physicians! Music is useful, too, for strengthening our mental energy and ideas, refreshing our imagination, reviving our sensitiveness and relieving fatigue. It enables the brain to continue a work, after exertion seems impossible. Many an artist and man of letters finds relaxation from his toil in this art.

One of my friends, a skilful painter, sits down to the piano to seek the inspiration which for the moment is denied him.

One of our most renowned poets plays the violin or violoncello, for the sake of arousing his poetic enthusiasm.

Our *savants*, *literati*, artists and sculptors, are

generally lovers of music, and this accounts for their being found in great numbers in theatres and concert-halls.

Our soldiers when on a campaign often hire their drummers to strike up a march when, wearied with fatigue, they come to a halt.

Very often in Italy, during the harvest season, I have seen men and women dance the *tarantelle* or *saltarello* for hours together, with the greatest energy, or rather most excitedly, to the music of a tambourine and mandolin. Apparently unmindful of fatigue, when listening to the instruments, their ardor seemed constantly fed by song and gesture; they were more like persons possessed by demons than laborers accustomed to bend under the hardest kind of labor for fifteen hours a day.

A propos of these fanatic Italian dances, it would perhaps be proper to speak of that animal to which so many wonderful virtues are attributed: I refer to the tarantula. What has not been said and written about this spider? I am ignorant as to what gave rise to the common tradition about it. Is it not claimed that persons stung by the tarantula are forced, spite of themselves, to dance when they hear certain music? Well, I have made many inquiries and investigations. I have questioned at length the inhabitants of both city and country, from Milan to Venice, from Venice to Florence, from Florence to Rome, from Rome to Naples, from Naples to Palermo and Messina; I have pursued my investigations in every direction, and no one has been able to give me a satisfactory answer. Only among the inhabitants of the Roman Campagna have I found some evidences —now nearly effaced—of the strange peculiarity attributed to this insect, which is regarded by all the peasants as poisonous; but I repeat, in spite of all my research, not only have I never succeeded in finding a person who has been stung by a tarantula, but I have never even seen the insect itself, although a great number of the inhabitants described it to me.

All the facts I have just related, all the phenomena I have just mentioned, are worthy of being carefully considered; destined to serve as a foundation for the therapeutic edifice which is to be erected, they are surely of great importance; they merit the attention of students and physicians. To-day, especially, when minds engaged mostly in philosophical pursuits, strive to free themselves from the yoke of prejudice, every man should bring a stone to assist in constructing a lasting monument. We must advance, in the name of progress, and we must use the knowledge gained by numerous efforts and studious application, to extend the possibilities of the art of music.

If we judge of the inefficacy of a remedy by its singularity; if the physician, dreading ridicule, hesitates or shrinks before the opinion be it true or false, which the world may form with regard to him, I admit we must give up the use of music in the treatment of disease. If, on the other hand, a remedy seems useful and promotes the relief of physical suffering, the physician should brave satire, walk with head erect, remain deaf to clamor, and not neglect any means, whatever they may be, of snatching from death, or at least relieving from suffering, the individual who has placed supreme confidence in him.

CHAPTER VIII.

Application of music.—As a preventive and curative means. —Diseases for which it may be employed.—Precautions to be taken in using it.

FROM what we have already seen, and from the details that have been given throughout this work, we may readily perceive the medical, therapeutic and hygienic resources which the art of music, or rather sound (that is, the sonorous fluid), offer. We should certainly have reason to be astonished that this powerful and vigorous remedy has remained so long neglected, did we not know how difficult it is to apply it. On the part of the physician, it calls for the most scrupulous attention, skilful care, and the most exact knowledge of the strength of the remedy and the sensitiveness of the patient; for if the combination of sounds has, in some cases, a salutary effect, in others the resuls may be lamentably fatal.

In many brain diseases, for instance in inflammation of that organ, a style of music that is suggested without reflection, or rashly employed, might produce very serious consequences; it might even prove very dangerous, especially to excessively nervous persons.

Ten years ago, I witnessed a very curious fact. A tall beautiful girl, eighteen or nineteen years old, of an intensely nervous disposition, passed an evening at a house where I was staying. One of Beethoven's quartets was being performed. This work, played as it was immediately after some of Mozart's compositions, made a very vivid impression on the young girl. She at first had a feeling of restlessness, then of extreme fatigue, followed by nervous agitation. These symptoms, at first courageously endured, increased in intensity, and after a time, the girl, no longer able to contain herself, uttered a cry, and was seized with violent convulsions. Those present crowded about her, and she was carried into an adjoining room, when the convulsive movement ceased, but she fell into a state of complete catalepsy. All music was stopped; but, spite of all efforts for her relief, the paroxysm lasted more than two hours.

At this time my ideas were not directed toward investigating the influence of music on the animal economy, so that I did not derive from this peculiar occurrence the profit that I should have done, had it happened ten years later. At the present time I am convinced that it would have been possible to shorten the duration of this cataleptic attack, or entirely prevent it, if the musicians had been made to perform compositions of a different style from those that had just been heard.

If we would apply music to the treatment or relief of disease we must necessarily be acquainted with the patient's manner of life, his character, temperament, habits and passions. The physician, being enlightened on all these peculiarities, will select the most suitable airs, being careful as regards rhythm, set them in fitting keys, and adapt them to the proper instruments. He will have the musician, or musicians, play in a room separated from the sick-chamber, and the performance will begin with music familiar to the patient. If the latter be habitually morose, the concert must be opened with a melancholy composition, passing on, by imperceptible degrees, to other works of a less grave character, and finally reaching the liveliest melodies by the most skilfully managed transitions; but scrupulously avoid, under all circumstances, songs likely to keep the patient's mind in the same condition into which he has fallen, and do not let the concert last too long.

If this first concert does not have quite the hoped for effect, but even one opposed to that intended, do not hastily conclude that the remedy is bad, useless or hurtful, and do not utterly condemn it without a new trial. It may happen that the organs, being charged too much or too little with the musical fluid, may be exposed to too strong or too feeble oscillations, to preserve the equilibrium indispensable to health, as well as the free exercise of all the functions. A second, a third concert, may restore the parts constituting the economy, to a normal state, by stimulating or relaxing them to a proper degree, and thus the general harmony be reëstablished.

The choice of musical compositions, the proper moment for applying them, a correct appreciation of the patient's constitution, comprise the whole secret of this method of cure.

In mental and nervous diseases, especially among women, the success of the *musical treatment* will be all the more marked, if directed by a skilful hand.

In agitation and trouble of soul, in restlessness of spirit, in the perversion of intellectual and moral ideas and faculties, care and foresight should be used lest the evil be aggravated instead of cured. While wishing to inspire hope, do not arouse despair, and when trying to dissipate the clouds that overshadow the mind, be careful not to increase the darkness.

Order lively and vigorous music for a delicate, weak and nervous child ; infuse this curative fluid by degrees into his brain and organs, following given rules.

Let the pale, delicate young girl listen to a lively style of music, and never have played within her hearing any compositions of a melancholy or emotional character, or she will fall into a state of despondency fatal to health and happiness; possibly even catalepsy may be induced.

Those women who are prematurely developed, ardent, enthusiastic and passionate, need to hear concerts that are lively but not exciting, gentle but not sensual music, and sometimes even, tones breathing a sweet melancholy. We must exercise caution and prudence in treating these impassioned creatures. Especially in the case of young girls who are susceptible to the fascinating influences of music, it is better to repress their musical propensities than allow them to be developed too easily. Think of Grétry's daughters, who died at thirteen and fifteen years of age, weakened and broken down by the study and composition of works beyond their powers.

Those of a dull, sluggish nature should be gradually roused by means of powerful and impressive music; those of a nervous disposition must be soothed by sweet and tender melodies. Let the person troubled with a bilious temperament hear songs that are light, short and tinged with gayety. Try to excite merriment in hypochondriacs, to induce calmness in those suffering from hysteria, and divert the epileptics from all thought of their paroxysms and convulsions. Above all, carefully avoid recalling, by your melodies, those thoughts which naturally absorb the whole mind, but from which it should be entirely freed.

R. P. Kircher, author of a bulky and illogical work on music, thinks that every composition has its own peculiar time and place. He thinks, moreover, that in the choice of music for concerts we should have a due regard to seasons. In his opinion, the months of May, June, July, August, September and October, that is, the warmest months, are the most favorable for instrumental performances. The months of November, December, January, February, March and April, that is, the season of fog, rain and snow, are the least adapted to the development and intensity of sound. The winds that dry the air, he says, are those most favorable for the voice; moist winds tend to change the sounds. Do we not recognize in these observations, and in the conditions described as favorable or unfavorable to the development of the sonorous fluid, the same conditions that are well or ill adapted to the manifestation of the electric fluid? The more we reflect, and the more attentively we observe facts, the more proofs we gain of the analogy or similarity between the imponderable fluids, which, I again repeat, might well be called varied modifications of one and the same fluid.

Generally, the music of the ancients is regarded as having more power to work marvels and touch the soul than that of the moderns; perhaps it is so, but how are we to account then, for the supposed inferiority of the composers of our own day? Are they less learned than the ancients? Certainly not, the very opposite of that may be maintained. Their harmonic and orchestral combinations have been surpassed by those of no other age, and their operas contain undeniably fine passages; why, then, should they suffer this reproach? The reason may be readily discovered. The composers of ancient times, full of taste and simplicity, ignorant of harmony, or at least, having but a very confused idea of it, conceived of no music but simple melodies and songs. They produced their effects by searching the inmost recesses of their own hearts, and understood how to express their own thoughts and feelings with perfect artlessness and ease. Our modern composers, on the contrary, paying more attention to combinations of harmony than of melody, appealing to the intellect rather than the soul, seek to obtain dramatic effects in studying works on instrumentation and orchestration; and by *noise*, that is, the fanciful combination of sounds, they hope to give proof of originality and thus gain renown. Their combinations are very learned and ingenious, no doubt, but they are lacking in melody, which is obscured and buried, as it were, under heaps of notes. The song, choked and stifled, so to speak, is hardly perceptible through this harmony, and the effort of following it is so great that we finally become nervous and restless, and all enjoyment ceases.

Compare the songs of Pergolesi, Palestrina, Gluck, Grétry, and Mozart, with those of our best modern dramatic composers, and see if, in the moments of pathos and passion, of grief and frenzy, these great masters ever made use of strange orchestral combinations as accompaniments for their melodies, thus destroying their character, under the pretext of enhancing their brilliancy. Music of this nature, when it reaches an audience, finds them cold and indifferent.

When a warrior is at the point of death, he never wastes breath singing *runs;* and Iphigenia standing at the sacrificial altar, ought not to express her grief in chromatic scales and endless *fioriture.* Is it not a shocking misconception to have any one who has suffered a stab from a dagger, or taken poison, use his last breath for singing grace notes and trills? No, the dramatic interest and musical accentuation ought not to be sacrificed to the voice of a *prima donna*, or that of a *tenore assoluto.*

These observations seem to me true, because they are accepted by all persons of taste, and the successful revival of some operas of former times proves how delightful a simple music is, based on the study of human sentiments.

Like Proteus, imperceptible, assuming countless forms and colors, speaking the language of all passions, music penetrates the very depths of the soul, and searches the inmost recesses of the heart; seeks out the tenderest emotions and most hidden sentiments, and by a sort of despotic power is able either to arouse or calm, to brighten or extinguish them.

Substituting with incredible facility ideas diametrically opposed to those already in our mind, it forces our thoughts, at its own will, into whatever direction it pleases. It subjects us to strange metamorphoses, and makes us ascend and descend the scale of all the passions and emotions in its whole vast extent.

It pervades everything; nature is all music, all harmony. The sighing breeze, the rippling brook, the warbling bird, as well as the roar of the storm, the rushing of the tempest, or the reverberation of the waves on the sandy shore—all nature is to man a source of sweet and delightful, or of sad and terrible sensations.

What, then, is music? Is it, as Scudo says, a

presentiment, or is it a reminiscence of blessings once felt, or is it, possibly, the intuition of a happy fulfilment of our hopes?

Finite beings as we are, why are we not satisfied with the finite? Why, in the enjoyment of abundance and pleasure, do certain simple tones and combinations, heard perhaps afar off, make us start, and fill our minds with sad forebodings?

THE END.

www.ingramcontent.com/pod-product-compliance
Lightning Source LLC
LaVergne TN
LVHW050527100826
845148LV00002B/468